Robust Control in Power Systems

POWER ELECTRONICS AND POWER SYSTEMS

Series Editors
M. A. Pai and Alex Stankovic

Other books in the series:

APPLIED MATHEMATICS FOR RESTRUCTURED ELECTRIC POWER SYSTEMS:
Optimization, Control, and Computational Intelligence
Joe H. Chow, Felix F. Wu, James A. Momoh, ISBN 0-387-23470-5
HVDC and FACTS Controllers: *Applications of Static Converters in Power Systems*
Vijay K. Sood, ISBN 1-4020-7890-0
POWER QUALITY ENHANCEMENT USING CUSTOM POWER DEVICES
Arindam Ghosh, Gerard Ledwich, ISBN 1-4020-7180-9
COMPUTATIONAL METHODS FOR LARGE SPARSE POWER SYSTEMS ANALYSIS:
An Object Oriented Approach,
S. A. Soman, S. A. Khaparde, and Shubha Pandit, ISBN 0-7923-7591-2
OPERATION OF RESTRUCTURED POWER SYSTEMS
Kankar Bhattacharya, Math H.J. Bollen and Jaap E. Daalder, ISBN 0-7923-7397-9
TRANSIENT STABILITY OF POWER SYSTEMS: *A Unified Approach to Assessment and Control*
Mania Pavella, Damien Ernst and Daniel Ruiz-Vega, ISBN 0-7923-7963-2
MAINTENANCE SCHEDULING IN RESTRUCTURED POWER SYSTEMS
M. Shahidehpour and M. Marwali, ISBN: 0-7923-7872-5
POWER SYSTEM OSCILLATIONS
Graham Rogers, ISBN: 0-7923-7712-5
STATE ESTIMATION IN ELECTRIC POWER SYSTEMS: *A Generalized Approach*
A. Monticelli, ISBN: 0-7923-8519-5
COMPUTATIONAL AUCTION MECHANISMS FOR RESTRUCTURED POWER INDUSTRY OPERATIONS
Gerald B. Sheblé, ISBN: 0-7923-8475-X
ANALYSIS OF SUBSYNCHRONOUS RESONANCE IN POWER SYSTEMS
K.R. Padiyar, ISBN: 0-7923-8319-2
POWER SYSTEMS RESTRUCTURING: *Engineering and Economics*
Marija Ilic, Francisco Galiana, and Lester Fink, ISBN: 0-7923-8163-7
CRYOGENIC OPERATION OF SILICON POWER DEVICES
Ranbir Singh and B. Jayant Baliga, ISBN: 0-7923-8157-2
VOLTAGE STABILITY OF ELECTRIC POWER SYSTEMS
Thierry Van Cutsem and Costas Vournas, ISBN: 0-7923-8139-4
AUTOMATIC LEARNING TECHNIQUES IN POWER SYSTEMS,
Louis A. Wehenkel, ISBN: 0-7923-8068-1
ENERGY FUNCTION ANALYSIS FOR POWER SYSTEM STABILITY
M. A. Pai, ISBN: 0-7923-9035-0
ELECTROMAGNETIC MODELLING OF POWER ELECTRONIC CONVERTERS
J. A. Ferreira, ISBN: 0-7923-9034-2
SPOT PRICING OF ELECTRICITY
F. C. Schweppe, M. C. Caramanis, R. D.Tabors, R. E. Bohn, ISBN: 0-89838-260-2
THE FIELD ORIENTATION PRINCIPLE IN CONTROL OF INDUCTION MOTORS
Andrzej M. Trzynadlowski, ISBN: 0-7923-9420-8
FINITE ELEMENT ANALYSIS OF ELECTRICAL MACHINES
S. J. Salon, ISBN: 0-7923-9594-8

ROBUST CONTROL IN POWER SYSTEMS

BIKASH PAL
Imperial College London

BALARKO CHAUDHURI
Imperial College London

Bikash Pal
Imperial College London

Balarko Chaudhuri
Imperial College London

Robust Control in Power Systems

Library of Congress Cataloging-in-Publication Data

A C.I.P. Catalogue record for this book is available
from the Library of Congress.

ISBN 0-387-25949-X   e-ISBN 0-387-25950-3   Printed on acid-free paper.
ISBN 978-0387-25949-9

© 2005 Springer Science+Business Media, Inc.
All rights reserved. This work may not be translated or copied in whole or in part without
the written permission of the publisher (Springer Science+Business Media, Inc., 233 Spring
Street, New York, NY 10013, USA), except for brief excerpts in connection with reviews or
scholarly analysis. Use in connection with any form of information storage and retrieval,
electronic adaptation, computer software, or by similar or dissimilar methodology now
know or hereafter developed is forbidden.
The use in this publication of trade names, trademarks, service marks and similar terms,
even if the are not identified as such, is not to be taken as an expression of opinion as to
whether or not they are subject to proprietary rights.

Printed in the United States of America.

9 8 7 6 5 4 3 2 1   SPIN 11054511

springeronline.com

Dedicated to our Parents

Contents

Dedication	v
List of Figures	xiii
List of Tables	
Preface	xix
Acknowledgments	xxi
Foreword	xxiii

1. INTRODUCTION   1
   References   3

2. POWER SYSTEM OSCILLATIONS   5
   2.1 Introduction   5
   2.2 Nature of electromechanical oscillations   5
       2.2.1 Intraplant mode oscillations   5
       2.2.2 Local plant mode oscillations   5
       2.2.3 Interarea mode oscillations   6
       2.2.4 Control mode oscillations   7
       2.2.5 Torsional mode oscillations   7
   2.3 Role of Oscillations in Power Blackouts   8
       2.3.1 Oscillations in the WECC system   9
   2.4 Summary   11
   References   11

3. LINEAR CONTROL IN POWER SYSTEMS   13
   3.1 Introduction   13
   3.2 Linear system analysis tools in power systems   14
       3.2.1 Eigenvalue analysis   16
       3.2.2 Modal controllability, observability and residue   20

		3.2.3	Singular values and singular vectors	23

		3.2.3	Singular values and singular vectors	23
		3.2.4	$\mathcal{H}_\infty$ and $\mathcal{H}_2$ norm	25
		3.2.5	Hankel singular values and model reduction	27
		3.2.6	Stability, performance and robustness	31
		3.2.7	Control design specifications in power systems	34
	3.3	Summary		36
	References			36
4.	TEST SYSTEM MODEL			39
	4.1	Overview of the test system		40
	4.2	Models of different components		41
		4.2.1	Generators	41
		4.2.2	Excitation systems	43
		4.2.3	Network power flow model	44
	4.3	Modelling of FACTS devices		45
		4.3.1	Thyristor controlled series capacitor (TCSC)	45
		4.3.2	Static VAr compensator (SVC)	48
		4.3.3	Thyristor controlled phase angle regulator (TCPAR)	50
	4.4	Linearized system model		53
	4.5	Choice of remote signals		54
	4.6	Simplification of system model		56
	References			58
5.	POWER SYSTEM STABILIZERS			59
	5.1	Introduction		59
	5.2	Basic Concept of PSS		60
	5.3	Stabilizing signals for PSS		62
	5.4	Structure of PSS		63
	5.5	Methods of PSS design		66
		5.5.1	Damping torque approach	67
		5.5.2	Frequency response approach	68
		5.5.3	Eigenvalue and state-space approach	69
		5.5.4	Summary	74
	References			75
6.	MULTIPLE-MODEL ADAPTIVE CONTROL APPROACH			79
	6.1	Introduction		79
	6.2	Overview of MMAC strategy		80

	6.2.1	Calculation of probability: Bayesian approach	81
	6.2.2	Calculation of weights	82
6.3	Study system		83
6.4	Model bank		84
	6.4.1	4-machine, 2-area system	84
	6.4.2	16-machine, 5-area system	85
6.5	Control tuning and robustness testing		85
	6.5.1	4-machine, 2-area system	85
	6.5.2	16-machine, 5-area system	87
6.6	Test cases		89
	6.6.1	Test case I	89
	6.6.2	Test case II	90
6.7	Choice of convergence factor and artificial cut-off		90
6.8	Simulation results with a 4-machine, 2-area study system		91
	6.8.1	Test case I	92
	6.8.2	Test case II	94
6.9	Simulation results with a 16-machine, 5-area study system		95
	6.9.1	Test case I	95
	6.9.2	Test case IIa	98
	6.9.3	Test case IIb	100
6.10	Summary		101
References			102
7.	SIMULTANEOUS STABILIZATION		105
7.1	Eigen-Value-Distance Minimization		105
7.2	Robust pole-placement		110
7.3	Case study		111
7.4	Control design		111
7.5	Simulation results		112
7.6	Summary		112
References			114
8.	MIXED-SENSITIVITY APPROACH USING LMI		115
8.1	Introduction		115
8.2	$\mathcal{H}_\infty$ mixed-sensitivity formulation		116
8.3	Generalized $\mathcal{H}_\infty$ problem with pole-placement		117
8.4	Matrix inequality formulation		119

	8.5	Linearization of the matrix inequalities	121
	8.6	Case study	123
		8.6.1 Weight selection	123
		8.6.2 Control design	123
		8.6.3 Performance evaluation	127
		8.6.4 Simulation results	128
	8.7	Case study on sequential design	131
		8.7.1 Test system	131
		8.7.2 Control design	132
		8.7.3 Performance evaluation	132
		8.7.4 Simulation results	134
	8.8	Summary	134
	References	136
9.	NORMALIZED $\mathcal{H}_\infty$ LOOP-SHAPING USING LMI	139
	9.1	Introduction	139
	9.2	Design approach	140
		9.2.1 Loop-shaping	140
		9.2.2 Robust stabilization	142
	9.3	Case study	144
		9.3.1 Loop-shaping	144
		9.3.2 Control Design	145
		9.3.3 Simulation results	146
	9.4	Summary	148
	References	149
10.	$\mathcal{H}_\infty$ CONTROL FOR TIME-DELAYED SYSTEMS	151
	10.1	Introduction	151
	10.2	Smith predictor for time-delayed or dead-time systems: an overview	153
	10.3	Problem formulation using unified Smith predictor	156
	10.4	Case study	158
		10.4.1 Control design	159
		10.4.2 Performance evaluation	161
		10.4.3 Simulation results with TCSC	161
	10.5	Simulation results with SVC	164
	10.6	Summary	166
	References	168

A	16-machine, 5-area System Power Flow Data		171
B	16-machine, 5-area System Dynamic Data		177
C	Jacobian of the FACTS Power Injection		179
	C.1 Thyristor controlled series capacitor (TCSC)		179
		C.1.1 W.r.t state variables	179
		C.1.2 W.r.t algebraic variables	179
	C.2 Static VAr compensator (SVC)		180
		C.2.1 W.r.t state variables	180
		C.2.2 W.r.t algebraic variables	180
	C.3 Thyristor controlled phase angle regulator (TCPAR)		180
		C.3.1 W.r.t state variables	180
		C.3.2 W.r.t algebraic variables	181
D	Matlab Routine for Controller Design Using LMI Control Toolbox		183
E	Matlab Routine for Controller Design Using "$hinfmix$" Function		187
Index			189

List of Figures

1.1	Spontaneous Oscillations on the Pacific AC Intertie, August 2,1974;Source: CIGRE Technical Report 111 on Analysis and Control of Power System Oscillations, 1996	2
2.1	A typical example of local oscillation	6
2.2	A typical example of interarea oscillation	7
2.3	A typical example of torsional mode oscillation	8
3.1	Model system in page 732 of Kundur's book	15
3.2	maximum singular value response of the example system	27
3.3	Frequency response of the original and the reduced 4th order system	31
3.4	Frequency response of the original and the reduced 3rd order system	31
3.5	Frequency response of the original (365 states) and reduced (20 states) system using Krylov subspace based techniques	32
3.6	Frequency response of the original (365 states) and reduced (20 states) system: simplified down to 100 states using Krylov subspace based techniques and then to 20 states using balanced truncation	32
4.1	Sixteen machine five area study system with a FACTS device	40
4.2	Thyristor controlled series capacitor (TCSC) topology	46
4.3	Voltage source model of TCSC	46
4.4	Power injection model of TCSC	47
4.5	Small-signal dynamic model of TCSC	48
4.6	Static VAr compensator (SVC) topology	49
4.7	Small-signal dynamic model of SVC	50

4.8	Thyristor controlled phase angle regulator (TCPAR) topology	51
4.9	Power injection model of TCPAR	52
4.10	Small-signal dynamic model of TCPAR	53
4.11	Frequency response of original and simplified system with TCSC	57
5.1	Heffron-Phillips block diagram of single machine infinite bus model	60
5.2	A commonly used structure of PSS	63
5.3	Response of a typical SMIB system under disturbance	66
6.1	Schematic of MMAC strategy	81
6.2	Study system	83
6.3	Frequency response of original and simplified plant	86
6.4	Frequency response of the controller	87
6.5	Performance of conventional controllers	88
6.6	Robustness test for conventional controllers	88
6.7	Variation of the computed weights	91
6.8	Test case I : Variation of the weights corresponding to each model	92
6.9	Test case I : Dynamic response of the system	93
6.10	Test case I : Power flow between buses #10 and #9	93
6.11	Test case I : Response of the controller	94
6.12	Test case II : Variation of the weights corresponding to each model	95
6.13	Test case II : Dynamic response of the system	96
6.14	Test case II : Power flow between buses #10 and #9	96
6.15	Test case II : Response of the controller	97
6.16	Test Case I: Variation of weights	97
6.17	Test Case I: Dynamic response of the system	98
6.18	Test Case IIa: Variation of weights	99
6.19	Test Case IIa: Dynamic response of the system	99
6.20	Test Case IIb: Variation of weights	100
6.21	Test Case IIb: Dynamic response of the system	101
7.1	Closed-loop feedback configuration with negative feedback	106
7.2	Closed-loop feedback configuration for a 1-input, 3-output system with positive feedback	109
7.3	Dynamic response of the system	113
7.4	Dynamic response of the system	113

List of Figures

8.1	Mixed-sensitivity formulation	117
8.2	Generalized regulator set-up for mixed-sensitivity formulation	118
8.3	Conic sector region for pole-placement	120
8.4	Frequency response of the weighting filters	123
8.5	Frequency response of the full and reduced controller	125
8.6	Frequency response of sensitivity (S)	125
8.7	Frequency response of control times sensitivity (KS)	126
8.8	Dynamic response of the system	129
8.9	Dynamic response of the system	130
8.10	Dynamic response of the system	130
8.11	Sixteen machine five area study system with three FACTS devices	131
8.12	Dynamic response of the system	134
8.13	Percentage compensation of the TCSC	135
8.14	Output of the SVC	135
8.15	Phase angle of TCPAR	136
9.1	Loop-shaping design procedure	141
9.2	Normalized coprime factor robust stabilization problem	142
9.3	Frequency response of the pre-compensator	145
9.4	Frequency response of the reduced order original and shaped system	146
9.5	Dynamic response of the system	147
9.6	Dynamic response of the system	147
9.7	Dynamic response of the system	148
10.1	Control setup for dead-time systems	153
10.2	An equivalent representation of dead-time systems	153
10.3	Introduction of Smith predictor and delay block	154
10.4	Uniform delay in both paths	155
10.5	Smith predictor formulation	155
10.6	Unified Smith predictor	157
10.7	Control setup with mixed-sensitivity design formulation	159
10.8	Frequency response of the weighting filters	160
10.9	Frequency response of the full and reduced controller	161
10.10	Dynamic response of the system with TCSC installed; controller designed with 0.75 s delay	163
10.11	Output of the TCSC	164
10.12	Dynamic response of the system with a delay of 0.5 s	165

10.13	Dynamic response of the system; controller designed without considering delay	165
10.14	Dynamic response of the system with SVC; controller designed considering delay	166
10.15	Output of the SVC	166
10.16	Dynamic response of the system with a delay of 0.5 s	167
10.17	Dynamic response of the system with a delay of 1.0 s	167
10.18	Dynamic response of the system; controller designed without considering delay	168

List of Tables

3.1	Eigenvectors and normalized participation factors corresponding to local mode $-0.282 + j8.62$	19
3.2	Modal controllability, observability and residue corresponding to local mode $-0.282 + j8.62$	23
4.1	Inter-area modes of the test system with TCSC	54
4.2	Inter-area modes of the test system with SVC	54
4.3	Inter-area modes of the test system with TCPAR	55
4.4	Normalized residues for active power flow signals from different lines with TCSC installed in the system	55
4.5	Normalized residues for active power flow signals from different lines with SVC installed in the system	56
4.6	Normalized residues for active power flow signals from different lines with TCPAR installed in the system	57
6.1	Critical modes of oscillation of the study system	83
6.2	Operating conditions used in the model bank	84
6.3	Operating conditions used in the model bank	85
6.4	Closed-loop damping ratio of the inter-area mode for different models and controllers	87
7.1	Specified and achievable pole locations for the reduced closed-loop system	112
8.1	Damping ratios and frequencies of the inter-area modes	127
8.2	Damping ratios and frequencies of the critical inter-area modes at different levels of power flow between NETS and NYPS	128
8.3	Damping ratios and frequencies of the critical inter-area modes for different load models	128

8.4	Damping ratios and frequencies of inter-area modes with the controller for TCSC (Control loops for SVC and TCPAR open)	133
8.5	Damping ratios and frequencies of inter-area modes with the controllers for TCSC and SVC (Control loop for TCPAR open)	133
8.6	Damping ratios and frequencies of inter-area modes with the controllers for TCSC, SVC and TCPAR (All the control loops closed)	133
9.1	Damping ratios and frequencies of inter-area modes with	146
10.1	Damping ratios and frequencies of the inter-area modes	162
10.2	Damping ratios and frequencies of the critical inter-area modes at different levels of power flow between NETS and NYPS	162
10.3	Damping ratios and frequencies of the critical inter-area modes for different load models	162
A.1	Machine bus data	171
A.2	Load bus data	172
A.2(continued)		173
A.3	Line data	173
A.3(continued)		174
A.3(continued)		175
B.1	Machine data	177
B.1(continued)		178
B.2	DC excitation system data	178
B.3	Static excitation system and PSS data	178

Preface

The aim of this monograph is to make a comprehensive presentation of recent research into the application of linear robust control theory to the damping of inter-area oscillations in power systems with FACTS devices. The subject is introduced with an overview of the application of power system stabilizers (PSS) and their coordination as described in the existing literature.

The monograph is directed at engineers engaged in the research, design and development of power systems with particular concern for power system stability and a background knowledge in this area is assumed. Reference books that are particularly relevant are:

    Power System Stability and Control: Kundur 1994
    Power System Dynamics and Stability: Sauer and Pai 1998
    Multivariable Feedback Control: Skogestad and Postlethwaite 2001

Power System Oscillations by Rogers (2000) is also relevant as an introduction since it is focused on the application of PSS to damp local and inter-area oscillations.

A brief historical account of oscillatory behavior in power systems is given in chapter 2. The analytic tools that are commonly used in small-signal stability analysis are presented in chapter 3 and chapter 4 contains a description of the components participating in interarea oscillations including FACTS devices:

    Static VAr capacitors (SVC)
    Thyristor-controlled series capacitors (TCSC)
    Thyristor-controlled phase shifters (TCPS)

The system model which is used to test damping controller designs is described in chapter 4.

Chapter 5 provides an overview of power system stabilizers (PSS) in a power system. The intention is to develop understanding and requirement of control design through damping torque concepts initially on a single machine infinite bus (SMIB) system. The extension of damping torque to the multi-machine system and different ways to achieve gain and phase compensation circuits used for PSS are discussed in the later part of this chapter.

A multiple-model based controller design is given in chapter 6 and chapter 7. A probability weighted approach is used to integrate the action of several controllers to give multiple-model adaptive control (MMAC). In chapter 7, a robust pole-placement approach giving eigenvalue distance minimization is used. Both methods address the robustness of the control schemes.

The $\mathcal{H}_\infty$ norm optimization is central to the controller design approaches in chapter 8 through 11. In chapter 8, a standard weighted mixed sensitivity optimization is made and a suitable set of linear matrix inequalities (LMI) being obtained numerically. Minimum closed-loop damping is ensured by pole-placement being taken as an additional LMI constraint.

In chapter 9, a left-coprime factorization approach gives a centralized control structure for a properly-shaped open loop plant using a loop-shaping technique. Again the numerical solution is obtained through LMI.

The effect of signal transmission delay on damping control is considered in chapter 10. A weighted-mixed sensitivity approach to the design of the central control structure is extended to include a delay in the output signal. Predictor techniques have been used in the controller design to obtain an $\mathcal{H}_\infty$ controller.

In all the above designs, robust damping for varying power level and changing network topology is confirmed by eigenvalue analysis and time-domain non-linear simulations have been made to demonstrate the validity of the designs.

BIKASH PAL AND BALARKO CHAUDHURI

Acknowledgments

Much of the information and insight presented in this book were obtained from the research done by the authors over the last seven years. It is our pleasure to acknowledge the support we received from various sources.

The Commonwealth Scholarship Commission of the Association of Commonwealth Universities, the Engineering and Physical Research Council (EPSRC) in the UK, and ABB Corporate Research, Switzerland have funded the research at various stages.

Dr Tim C. Green has been inspiring us throughout this research and also during the writing of this book. Professor (emeritus) M. A. Pai of the University of Illinois at Urbana Champaign has provided constant encouragement throughout. Professor (emeritus) Brian J. Cory and Dr Donald Macdonald have suggested many changes to our first draft and thoroughly edited the entire manuscript before submission.

Dr Imad M. Jaimoukha, Dr Argyrios C. Zolotas and Dr Haitham El-Zobaidi have helped us in many occasions through their expertise in multi-variable control and model reduction. Mr. Rajat Majumder and Dr Q. -C. Zhong have helped us significantly in formulating predictor approach to time-delayed $\mathcal{H}_\infty$ control. Mr. Majumder has also carried out a significant part of the simulation work in Matlab Simulink to produce the results presented in this monograph. We thank Mr. Alex Green and his colleagues at Springer for their editorial help.

We are grateful to our families for their supports during our work on this monograph.

Foreword

Low frequency electromechanical oscillations, with frequencies ranging from 0.1 to 2 Hz, are inherent to electric power systems. Problems due to inadequate damping of such oscillations have been encountered throughout the history of power systems. The earliest problems, which were experienced in the 1920s, were in the form of spontaneous oscillations or hunting. These were solved by the use of damper windings in the generators and turbine-type prime movers with favorable torque speed characteristics.

As power systems evolved, they were operated ever closer to transient and small-signal rotor angle stability limits. System stability characteristics were largely influenced by the strength of the transmission network, and the lack of sufficient synchronizing torque was the principal cause of system instability. The application of continuously acting voltage regulators contributed to the improvement in small-signal (or steady-state) stability. In the 1950s and 1960s, utilities were primarily concerned with transient stability. However, this situation has gradually changed since the late 1960s. Significant improvements in transient stability performance have been achieved through the use of high response exciters and special stability aids.

The above trends have been accompanied by an increased tendency of power systems to exhibit oscillatory instability. High response exciters, while improving transient stability, adversely affect the damping of local plant modes of oscillation, which have frequencies ranging from 0.8 to 2 Hz. The effects of fast exciters are compounded by the decreasing strength of transmission network relative to the size of generating stations. Adequate damping of local plant mode oscillations can be readily achieved by using power system stabilizers to modulate generator excitation controls.

Another source of oscillatory instability has been the formation of large groups of loosely coupled machines connected by weak links. This situation has developed as a consequence of growth in interconnections among power systems. With heavy power transfers, such systems exhibit inter-area modes of oscillation of low frequency. The stability of these modes has become a source of concern in today's power systems. There have been many reported occurrences of poorly damped or unstable inter-area oscillations. In some cases, this form of oscillatory instability has been the cause of major system blackouts. Large interconnected power systems usually exhibit several dominant modes of inter-area oscillations with frequencies ranging from 0.1 Hz to 0.8 Hz.

The use of supplementary controls is generally the only practical method of mitigating inter-area oscillation problems, without resorting to costly operating restrictions or transmission reinforcements. A number of power system devices have the potential for contributing to the damping of the oscillations by supplemental control. The use of power system stabilizers to control excitation of generators is often the most cost-effective method. The controllability of the inter-area modes of oscillation through excitation control is a function of many factors: location of the generator in relation to the oscillation mode shape, size and characteristics of nearby loads, and types of exciters on other nearby generators.

Supplemental stabilizing signals may also be used to control HVDC transmission links and SVCs to enhance damping of inter-area oscillations, depending on their location. While these devices are installed based primarily on other system considerations, their potential for controlling poorly damped system oscillations are often taken advantage of; many HVDC transmission and SVC installations are equipped with special modulation controls to stabilize inter-area oscillations.

In recent years, there has been considerable interest in the application of power electronic devices for enhancing the controllability, and hence the power transfer capability, of ac transmission; this concept is referred to as "FACTS" (Flexible AC Transmission System). The FACTS devices can provide fast continuous control of power flow in the transmission system by controlling voltages at critical buses, by changing the impedance of transmission lines, or by controlling the phase angles between the ends of transmission lines. This is an extension of the concept used by SVCs for enhancing transmission system capacity by rapid control of bus voltages. Apart from the SVC, two FACTS devices that can be effectively used for damping of system oscillations are the thyristor controlled series capacitor (TCSC) and the thyristor controlled phase angle regulator (TCPAR). FACTS devices, depending on the power system configuration and nature of the inter-area oscillations, may offer the most economic means of mitigating the problems.

A number of approaches and techniques are available for the design of controls for damping of inter-area oscillations. One important issue in the design and performance of the controllers is robustness. The controller should perform the desired function over the wide range of conditions encountered in the operation of the power system. Also, since equipment models are known only approximately, the controller must have minimum sensitivity to parameters of systems elements.

The gain and phase compensation approach in classical control theory has by far been the most effective and widely adopted method of designing power system controls. However, when there are a large number of control loops, a coordinated approach is necessary to address the variation in plant models for a range of operating conditions. The multivariable approach to control analysis and synthesis, under such situations, could provide a systematic approach to quantify the measure of robustness of prescribed control action.

This monograph on Robust Control in Power Systems deals with the applications of new techniques in linear system theory to control electromechanical oscillations in power systems. The primary consideration is to maintain adequate damping of inter-area oscillations under varying operating scenarios. Dr. Bikash Pal and Dr. Balarko Chaudhuri have explored the supplementary control through various FACTS controllers such as SVC, TCSC and TCPAR.

An overview of linear system theory from the perspective of power system damping control has been explained through some examples. The damping control design has been formulated as a norm optimization problem. The $\mathcal{H}_\infty$, $\mathcal{H}_2$ norm of properly defined transfer functions are minimized in the linear matrix inequalities (LMI) framework to obtain desired performance and stability robustness. Both centralized and decentralized control structure have been used. With the wide area control systems (WACS) through synchronized phasor measurement units (PMUs) in place, various control structures and solutions suggested here will be useful for secure operation of interconnected power systems.

Considering the heavily stressed state of the present day power systems, robust controls will play a major role in their secure and reliable operation. I strongly believe this monograph will be an invaluable source of reference on the subject.

Prabha S. Kundur
President & CEO Powertech Labs Inc.

Surrey, B.C. Canada
February 2005

Chapter 1

INTRODUCTION

Electromechanical oscillations appeared as soon as synchronous generators were operated in parallel. The mechanical inertia and power angle characteristics led to oscillations of 1-3 Hz and were described as *'hunting'*. Wagner observed that hydro generators always hunted at light load [Wagner, 1931]. High governor gain produced little damping and fluctuating load gave rise to continuous mechanical oscillations [Concordia, 1969]. Pole face damper windings were first suggested in Germany [Dreyfus, 1911] but as they gave rise to higher fault currents, they did not immediately find favour. However the benefits of damping in disturbed conditions were eventually recognized and pole face windings came into general use [Crary and Duncan, 1941].

As power systems became interconnected, areas of generation were found to be prone to oscillations at 0.2-1.0 Hz [Paserba, 1996]. The use of high gain voltage regulation in order to improve first swing transient stability exacerbated the oscillations. A proliferation of automatic controls were suggested with feed back of a mixed-bag of signals whose action was (and still is) difficult to understand. As the level of power transmission rose, largely through existing interconnections which were becoming weak and inadequate, load characteristics added to the problem causing spontaneous oscillations at particular times of the day as shown in Fig. 1.1 [Paserba, 1996]. The system was first-swing stable but unstable with growing oscillation.

Rotor damping circuits in pole face windings and solid rotors had a helpful effect in damping single machine oscillations but contributed little at the lower frequencies of interarea oscillations , damping currents penetrating to greater depths and giving lower damping circuit resistance. Additionally Kron showed that damping is inversely proportional to the square of the effective impedance

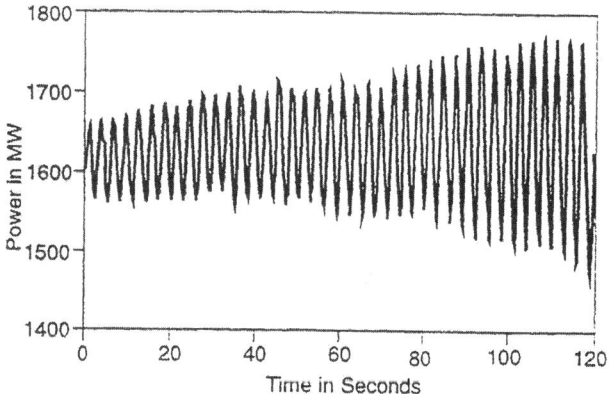

Figure 1.1. Spontaneous Oscillations on the Pacific AC Intertie, August 2,1974;Source: CIGRE Technical Report 111 on Analysis and Control of Power System Oscillations, 1996

at oscillation frequency between adjacent areas [Kron, 1952]. An expression for damping torque for a single-machine-infinite busbar system [SMIB] is given by Pai and others with similar conclusions [Pai et al., 2004].
Prior to 1960 multi-machine dynamic analysis was not available and the SMIB model was considered representative of all interarea conditions. Classical control theory was used for designing controllers and PSS were suggested with speed signals being fed to voltage regulators [de Mello and Concordia, 1969]. This proved very successful where the SMIB representations of interarea effects were adequate and resulted in improved damping [Larsen and Swann, 1981]

FACTS devices, static capacitors, controllable series capacitors and phase shifters, can provide fast control of network voltage and power flow and contribute towards the damping of interarea oscillations and enhance power transfer capacity [Hingorani and Gyugyi, 2000, Song and Johns, 1999]. However small signal stabilityis not always adequate and a supplementary strategy may be necessary.

Large power systems with many generating stations operating together contain a multitude of control circuits each with its own frequency characteristics. Adverse interactions may occur [Martins, 2000]. Rogers shows the detrimental effect of an intraplant mode with a poorly designed PSS [Rogers, 2000]. Generators tripped because of AVR and underexcitation limiters [Choi et al., 1993]. Interactions between HVDC controls and the ac system in western USA have occurred for many years [Martins, 2000]. Voltage instability leading to system collapse has been caused by the actions of an on-load tap changer (OLTC) controller [Cutsem and Vournas, 1998].

The multivariable approach enables the full character of a power system to be addressed. The multiple-input-multiple-output system (MIMO) offers the possibility of including the effect of all the interactions in the model representation. The adverse interactions may be seen as a lack of stability and performance margin. The limitations incurred when a multivariable system is designed as a series of SISO representations was been shown by Skogestad and Postlethwaite [Skogestad and Postlethwaite, 2001]. The singular value decomposition (SVD) approach allows for multi-variable directionality to be assessed. The maximum singular value can be seen as the worst gain in the worst direction, indicating an interaction which is a potential hazard.

Bode's stability criterion for a SISO system is not applicable to a MIMO analysis information. The design is consequently rather conservative but provides for the minimization of adverse interaction, cross-coupling terms being fully represented in the design process.

The power system design problem involves providing satisfactory control for varying demand and system topology. In the multivariable approach, plant model uncertainty is represented through a transfer function of finite norm as a bound of plant perturbation in the frequency domain. The norm-based design addresses robust stability and performance. The issues are addressed here as:

- General analysis and assessment of the factors influencing interarea oscillations
- The role and dynamic models of FACTS devices
- The stand alone and coordinated design of PSS in single and multi-machine systems
- Robust and multivariable control design techniques to mitigate oscillations through FACTS devices such as static VAr Compensator (SVC), thyristor controlled series capacitor (TCSC) and thyristor control phase shifter(TCPS).
- Techniques for addressing delays incurred in transmitting signals from remote locations for use in central controllers.

References

[Choi et al., 1993] Choi, S.S., Larkin, R., Bastik, M.T., and Ferres, A.W. (1993). Effects of underexcitation limiters on operations of remote generating station. *IEE Proceedings on Generation Transmission and Distribution*, 140(03):221–228.

[Concordia, 1969] Concordia, Charles (1969). Effect of prime-mover speed control characteristics on electric power system performance. *IEEE Transactions on Power Apparatus and Systems*, 88(5):752–756.

[Crary and Duncan, 1941] Crary, S.B. and Duncan, W.E. (1941). Amortisseur windings for hydrogenerators. *Electrical World*, 115:2204–2206.

[Cutsem and Vournas, 1998] Cutsem, T Van and Vournas, C (1998). *Voltage Stability of Electric Power Systems*. Kluwer Academic Press, USA.

[de Mello and Concordia, 1969] de Mello, F.P. and Concordia, Charles (1969). Concepts of synchronous machine stability as affected by excitation control. *IEEE Transactions on Power Apparatus and Systems*, 88(5):317–329.

[Dreyfus, 1911] Dreyfus, L. (1911). Einfuhrung in die theorie der selbsterregten schwmgungen synchroner maschinen. *Elektrotechnik und Maschinenbau*, 29:323–329.

[Hingorani and Gyugyi, 2000] Hingorani, N.G. and Gyugyi, L. (2000). *Understanding FACTS*. IEEE Press, USA.

[Kron, 1952] Kron, G. (1952). A new theory of hunting. *A., THAT IS,E Transactions*, 71:859–866.

[Larsen and Swann, 1981] Larsen, E.V. and Swann, D.A. (1981). Applying power system stabilizers, part i, ii and iii. *IEEE Transactions on Power Apparatus and Systems*, PAS-100(6):3017–3046.

[Martins, 2000] Martins, N. (2000). Impact of interactions among power system controls. *CIGRE Special Publication 38.02.16*, Technical Brochure 166.

[Pai et al., 2004] Pai, M.A., Sengupta, D.P., and Padiyar, K.R. (2004). *Topics in Small Signal Analysis of Power Systems*. Narosa Publishing House Private Limited, India.

[Paserba, 1996] Paserba, J. (1996). Analysis and control of power system oscillation. *CIGRE Special Publication 38.01.07*, Technical Brochure 111.

[Rogers, 2000] Rogers, G.J. (2000). The application of power system stabilizers to a multi-generator plant. *IEEE Transactions on Power Systems*, 15(1):350–355.

[Skogestad and Postlethwaite, 2001] Skogestad, S. and Postlethwaite, I. (2001). *Multivariable Feedback Control*. John Wiley and Sons, UK.

[Song and Johns, 1999] Song, Y.H. and Johns, A.T. (1999). *Flexible AC Transmission Systems*. IEE Power and Energy series, UK.

[Wagner, 1931] Wagner, C.F. (1931). Damper winding for water wheel generators. *American Institutions of Electrical Engineers*, 50:140–151.

Chapter 2

POWER SYSTEM OSCILLATIONS

2.1 Introduction

Oscillations in power systems are classified by the system components that they effect. Some of the major system collapses attributed to oscillations are described.

2.2 Nature of electromechanical oscillations

Electromechanical oscillations are of the following types:

- Intraplant mode oscillations
- Local plant mode oscillations
- Interarea mode oscillations
- Control mode oscillations
- Torsional modes between rotating plant

2.2.1 Intraplant mode oscillations

Machines on the same power generation site oscillate against each other at 2.0 to 3.0 Hz depending on the unit ratings and the reactance connecting them. This oscillation is termed as *intraplant* because the oscillations manifest themselves within the generation plant complex. The rest of the system is unaffected.

2.2.2 Local plant mode oscillations

In local mode, one generator swings against the rest of the system at 1.0 to 2.0 Hz. The variation in speed of a generator is shown in Fig. 2.1.

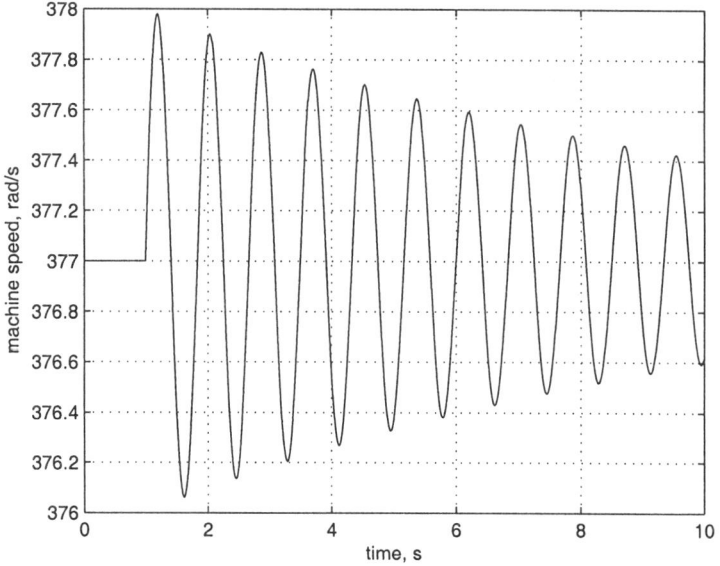

Figure 2.1. A typical example of local oscillation

The impact of the oscillation is localized to the generator and the line connecting it to the grid. The rest of the system is normally modelled as a constant voltage source whose frequency is assumed to remain constant. This is known as the single-machine-infinite-bus (SMIB) model. The damping and frequency vary with machine output and the impedance between the machine terminal and the infinite bus voltage. The oscillation may be removed with a single or dual input PSS that provides modulation of the voltage reference of the automatic voltage regulator (AVR) with proper phase and gain compensation circuit [Lee, 1992].

2.2.3 Interarea mode oscillations

This phenomenon is observed over a large part of the network. It involves two coherent group groups of generators swinging against each other at 1 Hz or less. The variation in tie-line power can be large as shown in Fig. 2.2. The oscillation frequency is approximately 0.3 Hz.

This complex phenomenon involves many parts of the system with highly non-linear dynamic behavior. The damping characteristic of the interarea mode is dictated by the tie-line strength, the nature of the loads and the power flow

Power System Oscillations 7

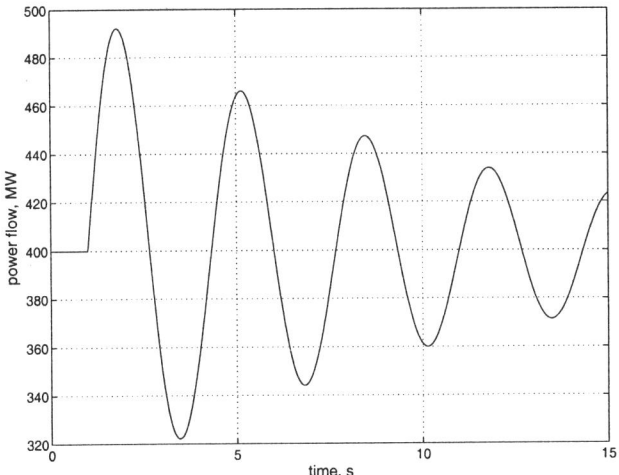

Figure 2.2. A typical example of interarea oscillation

through the interconnection and the interaction of loads with the dynamics of generators and their associated controls. The operation of the system in the presence of a lightly damped interarea mode is very difficult.

2.2.4 Control mode oscillations

These are associated with generators and poorly tuned exciters, governors, HVDC converters and SVC controls. Loads and excitation systems can interact through control modes [Rajagopalan et al., 1992]. Transformer tap-changing controls can also interact in a complex manner with non-linear loads giving rise to voltage oscillations [Cutsem and Vournas, 1998].

2.2.5 Torsional mode oscillations

These modes are associated with a turbine generator shaft system in the frequency range of 10-46 Hz. A typical oscillation is shown in Fig. 2.3.

Usually these modes are excited when a multi-stage turbine generator is connected to the grid system through a series compensated line [Padiyar, 1999]. A mechanical torsional mode of the shaft system interacts with the series capacitor at the natural frequency of the electrical network. The shaft resonance appears when network natural frequency equals synchronous frequency minus torsional frequency.

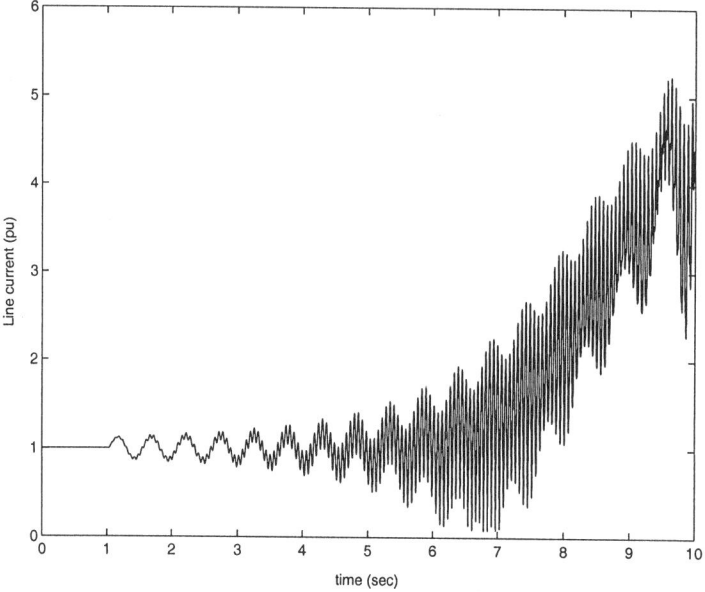

Figure 2.3. A typical example of torsional mode oscillation

2.3 Role of Oscillations in Power Blackouts

Interarea oscillations have led to many system separations but few wide-scale blackouts [Pal, 1999, Paserba, 1996]. Note worthy incidents include:

- Detroit Edison (DE-Ontario Hydro (OH)-Hydro Quebec (HQ) (1960s, 1985)
- Finland-Sweden-Norway-Denmark (1960s)
- Saskatchewan-Manitoba Hydro-Western Ontario (1966)
- Italy-Yugoslavia-Austria (1971-1974)
- Western Electric Coordinating Council (WECC) (1964,1996)
- Mid-continent area power pool (MAPP) (1971,1972)
- South East Australia (1975)
- Scotland-England (1978)
- Western Australia (1982,1983)
- Taiwan (1985)
- Ghana-Ivory Coast (1985)

- Southern Brazil (1975-1980,1984)

The power blackout of August 10, 1996 in the Western Electricity Co-ordination Council (WECC) (formerly WSCC) area is described below. It indicates the importance of understanding and managing oscillations for secure operation of the grid.

2.3.1 Oscillations in the WECC system

Power transfer capability in this system has been limited by stability considerations for 40 years because of the long distance between load centers and power sources. Oscillations have resulted in system separation on several occasions. They were caused by insufficient damping and synchronizing torque. The history of interarea oscillations in this system has influenced the system planning, design and operation strategy. Insufficient damping turned out to be the major constraint when in 1964, the Northwest United States and Southwest United States were interconnected through the Colorado River Storage Project. In less than a year of interconnected operation, there were at least a hundred tie-line separations due to system oscillations of power, frequency and voltage. In 1965, the problem was solved by modifications to one of the hydro-unit governors [Schleif et al., 1967].

About that time work was initiated to develop time domain stability programs for more detailed analysis of interconnected systems. This was very useful since it coincided with the planning of many 345 kV and 500 kV transmission projects, including the Northwest-Southwest Inter-tie which consisted of two 500 kV ac lines and ± 400 kV dc circuits. The initial plan was to carry 2000 MW through the ac circuits and 1440 MW through the dc line. Stability performance assessment showed that there was insufficient damping torque for ac power flows exceeding 1300 MW. It was found from the study that undamped oscillations of power, frequency and voltage at about 0.33 Hz was the major restraint on a larger transfer [Schleif and White, 1966]. It was later realized that many of the generator high gain automatic voltage regulators (AVR) produced negative damping at around 0.33 Hz which led to the development and application of PSS. It was found from the time domain simulations that there would be sufficient damping for the most severe disturbance with 1800 MW transferred through the ac lines if all generators in the system were equipped with PSS. After all the units were retrofitted with PSS, the oscillations disappeared and the stability limit depended only upon the synchronizing torque.

The Bonneville Power Authority (BPA) implemented a 1400 MW braking resistor at Chief Joseph Dam in 1974 to improve first swing stability of the system. This indicated that the system could operate with up to 2500 MW flowing through the AC interconnection with adequate stability margin following a severe disturbances such as a close in three phase fault. With even higher loading, however, slowly growing oscillations were observed, indicating that insufficient damping torque was again a problem at the higher loading level. The problem was relieved by the development of a scheme [Cresap et al., 1978] to modulate the northern terminal of the Northwest-Southwest dc line in such a manner as to provide positive damping to the ac system at the inter-tie frequency.

Overall the transmission capacity was increased from 1300 MW to 2500 MW without adding any transmission circuits. The only system additions were PSS, braking resistors and HVDC modulation. Many other interfaces in western USA are limited by insufficient damping torque and are highly dependent on PSS and other devices to provide positive damping. Currently there is a 0.7 Hz lightly damped interarea mode identified from system models and analytical techniques. In one interface, nearly 750 Mvar of static VAr compensators have been installed to add damping so that the full planned transmission capacity will be available [Lee et al., 1994].

On August 10, 1996, the Pacific AC inter-tie (PACI) emerged from the dormant state that had lasted since 1974 when the entire inter-connected system split into four islands with the loss of approximately 30 GW of load. More than 7 million customers were affected by this catastrophic event [Kosterev et al., 1999]. The mechanism of failure was a transient oscillation, under conditions of high power transfer on long paths that had been progressively weakened through a series of fairly routine resource losses. This series of events was simulated based on the dynamic model data base with data assembled from the data bases of the utilities. The simulation showed a well damped response for the critical set of contingencies but did not show any voltage decay. The power flow through the pacific HVDC tie was observed constant because of constant power control in the simulation model. The simulated frequency dip was also only 60% of the recorded value. On the other hand, undamped oscillations in the inter-tie power flow were recorded whilst voltages at several locations were depressed. Also the power flow through the HVDC tie was observed to vary thereby showing a serious discrepancy between the simulation model and the actual system dynamic characteristics. The oversimplified model of the HVDC tie and its control were replaced with four-terminal links and control at converter levels. The automatic governor control (AGC) was included during the transient which is normally omitted from dynamic simulations. The presence of large turbo-generators delayed the power output pick-up immediately following a frequency decay. This was done by not representing the governor action for large units. With all these modifications, the simulated system response differed appreciably from the recorded observation until a dynamic load model was included.

2.4 Summary

The long history of interarea oscillations in the WECC system and other interconnected systems, [Paserba, 1996], clearly identifies inadequate damping as the primary factor leading to system separation. The amount of damping and the frequency of oscillation varies with system operating conditions. The operating range of a power system is usually very wide, requiring an oscillation damping control strategy that is effective over this whole range. It is necessary to have comprehensive modelling and analysis techniques of all the components that may interact to produce oscillations.

References

[Cresap et al., 1978] Cresap, R.L., Scott, D.N., Mittelstadt, W.A., and Taylor, C.W. (1978). Damping of pacific ac intertie oscillations via modulation of the parallel pacific hvdc intertie. *CIGRE Paper*, 14(5).

[Cutsem and Vournas, 1998] Cutsem, T Van and Vournas, C (1998). *Voltage Stability of Electric Power Systems*. Kluwer Academic Press, USA.

[Kosterev et al., 1999] Kosterev, D.N., Taylor, C.W., and Mittelstadt, W. (1999). Model validation for the august 10, 1996 wscc system outage. *IEEE Transactions on Power Systems*, 14(3):967–979.

[Lee, 1992] Lee, D.C. (1992). *IEEE recommended practice for excitation system models for power system stability studies*. Energy development and power generation committee of power engineering society.

[Lee et al., 1994] Lee, R.L., Beshir, M.J., Finely, A.T., Hayes, D.R., Hsu, J.C., Peterson, H.R., Deshazo, C.L., and Gerlach, D.W. (1994). Application of static var compensator for the dynamic performance of the mead-adelanto and mead-phoenix transmission projects. *IEEE PES T and D Conference and Exposition, Chicago,IL*.

[Padiyar, 1999] Padiyar, K.R. (1999). *Analysis of Subsynchronous Resonance in Power Systems*. Kluwer Academic Publishers, USA.

[Pal, 1999] Pal, B.C. (1999). *Robust Damping Control of Inter-area Oscillations in Power System with Super-conducting Magnetic Energy Storage Devices*. PhD thesis, Imperial College of Science Technology and Medicine, Department of Electrical and Electronic Engineering.

[Paserba, 1996] Paserba, J. (1996). Analysis and control of power system oscillation. *CIGRE Special Publication 38.01.07*, Technical Brochure 111.

[Rajagopalan et al., 1992] Rajagopalan, C., Lesieutre, B., Sauer, P.W., and Pai, M.A. (1992). Dynamic aspects of voltage/power characteristics in multi-machine power systems. *IEEE Transactions on Power Systems*, 7(3):990–1000.

[Schleif et al., 1967] Schleif, F.R., Martin, G.E., and Angell, R.R. (1967). Damping of system oscillations with a hydro-generating unit. *IEEE Transactions on Power Apparatus and Systems*, 86(4):438–442.

[Schleif and White, 1966] Schleif, F.R. and White, J.H. (1966). Damping for northwest-southwest tie line oscillations - an analogue study. *IEEE Transactions on Power Apparatus and Systems*, 85(12):1234–1247.

Chapter 3

LINEAR CONTROL IN POWER SYSTEMS

3.1 Introduction

Many important components in a power system such as generators, excitation systems, governors and loads have very non-linear characteristics. These components and their associated controls include saturation and output limitations. The theory of nonlinear systems can be used to analyze these nonlinearities, however, the application is restricted to small and simple systems. The concept of energy functions has been applied as a powerful tool [Pai, 1989, Pavella and Murthy, 1994, Fouad and Vittal, 1992] to assess system security, stability limits and region of attraction of the post-fault equilibrium state. Suitable energy-like functions are constructed and examined to see whether their values diminish with time in the post-disturbance period. The construction of an energy function is easy as long as the classical generator model is considered with constant impedance loads. In the presence of larger order model complexities such as excitation control, turbine control, dynamic load and a network with transfer conductances etc., suitable energy functions are difficult to obtain.

However, the theory of linear system analysis can provide useful insight into the operating behavior of an interconnected power system although the dynamic behavior of the system must be assumed linear for such tools to be applicable. Fortunately, low frequency oscillations in a power system are fairly linear when caused by small disturbances such as the random fluctuation of generation and load. The variations in system dynamic variables such as machine angle and speed are also small under these conditions and the assumption of a linear system model around an operating equilibrium has provided valuable results. More often than not the conclusions drawn are consistent with what is observed in the field under similar operating circumstances. A better understanding of the nature of the system dynamics helps to plan control strategies for secure operation.

A brief overview of the tools of linear system theory used in this book for analysis of power systems and for their robust control synthesis is given below. The linear state space model of a sample power system is used to indicate the meaning of the various terms and definitions in the context of the dynamics, control and stability of power systems.

3.2 Linear system analysis tools in power systems

Low frequency electromechanical oscillations range from less than 1 Hz to 3 Hz other than those with sub-synchronous resonance (SSR). Multi-machine power system dynamic behavior in this frequency range is usually expressed as a set of non-linear differential and algebraic (DAE) equations. The algebraic equations result from the network power balance and generator stator current equations. The high frequency network and stator transients are usually ignored when the analysis is focused on low frequency electromechanical oscillations. The initial operating state of the algebraic variables such as bus voltages and angles are obtained through a standard power flow solution. The initial values of the dynamic variables are obtained by solving the differential equations through simple substitution of algebraic variables into the set of differential equations. The set of DAE is then linearized around the equilibrium point and a set of linear DAE is obtained:

$$\dot{x} = f(x, z, u) \qquad (3.1)$$
$$0 = g(x, z, u) \qquad (3.2)$$
$$y = h(x, z, u) \qquad (3.3)$$

where f and g are vectors of differential and algebraic equations and h is a vector of output equations. The inputs are normally reference values such as speed and voltage at individual units and can be voltage, reactance and power flow as set in FACTS devices. The output can be unit power output, bus frequency, bus voltage, line power or current etc. The notation $x \in R^n, z \in R^m, u \in R^p$ and $y \in R^q$ denotes the vectors of state variables, algebraic variables, inputs and outputs respectively. Linearizing (3.1) to (3.3) around the equilibrium point $\{x_0, z_0, u_0\}$ gives the following equations (3.4) to (3.6)

$$\Delta \dot{x} = \frac{\partial f}{\partial x} \Delta x + \frac{\partial f}{\partial z} \Delta z + \frac{\partial f}{\partial u} \Delta u \qquad (3.4)$$

$$0 = \frac{\partial g}{\partial x} \Delta x + \frac{\partial g}{\partial z} \Delta z + \frac{\partial g}{\partial u} \Delta u \qquad (3.5)$$

$$\Delta y = \frac{\partial h}{\partial x} \Delta x + \frac{\partial h}{\partial z} \Delta z + \frac{\partial h}{\partial u} \Delta u \qquad (3.6)$$

Elimination of the vector algebraic variable Δz from (3.4) and (3.6), gives

$$\Delta \dot{x} = A \Delta x + B \Delta u \tag{3.7}$$
$$\Delta y = C \Delta x + D \Delta u \tag{3.8}$$

where A, B, C, D are the matrix of partial derivatives in (3.4) to (3.6) evaluated at equilibrium $\{x_0, z_0, u_0\}$ as follows:

$$A = \left[\frac{\partial f}{\partial x} - \frac{\partial f}{\partial z} \left(\frac{\partial g}{\partial z} \right)^{-1} \frac{\partial g}{\partial x} \right], B = \left[\frac{\partial f}{\partial u} - \frac{\partial f}{\partial z} \left(\frac{\partial g}{\partial z} \right)^{-1} \frac{\partial g}{\partial u} \right] \tag{3.9}$$

$$C = \left[\frac{\partial h}{\partial x} - \frac{\partial h}{\partial z} \left(\frac{\partial g}{\partial z} \right)^{-1} \frac{\partial g}{\partial x} \right], D = \left[\frac{\partial h}{\partial u} - \frac{\partial h}{\partial z} \left(\frac{\partial g}{\partial z} \right)^{-1} \frac{\partial g}{\partial u} \right] \tag{3.10}$$

Power system state space representation is normally linearized around an operating point (hence the term small signal). The symbol Δ from (3.7) and (3.8) is omitted so as to follow the standard state space making x and u into the incremental values. This is the representation of a linearized DAE model of a power system on which standard linear analysis tools can be applied.

The SMIB system shown in Fig 3.1 is taken from page 732 of [Kundur, 1994] and is used here for illustration. Four rotor windings in the rotor(field and three dampers) are assumed. The excitation system is of high gain and is fast-acting. A static var compensator (SVC) is connected at bus 2. The model of the SVC is also described in Chapter 17 of [Kundur, 1994]. The real power in the line between bus 2 and 3 is used as stabilizing signal.

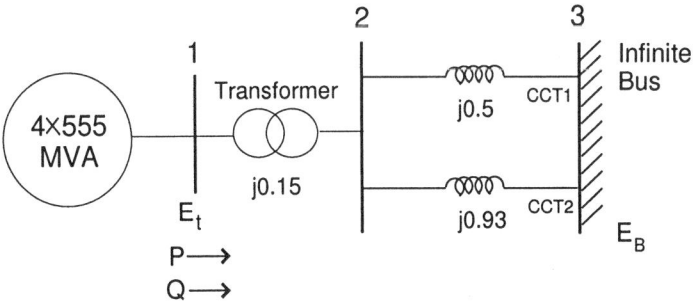

Figure 3.1. Model system in page 732 of Kundur's book

3.2.1 Eigenvalue analysis

Eigenvalues of a matrix A is found with the help of *eig(A)* function available in *Matlab* [mat, 1998]. In the example system, the function produced the eigenvalues as:

$$\Lambda = \begin{bmatrix} -94.66 \\ -49.907 \pm j4.2148 \\ -0.28191 \pm j8.6251 \\ -17.621 \\ -30.159 \\ -26.31 \\ -3.9119 \\ -0.10432 \end{bmatrix} \quad (3.11)$$

6 of the 10 eigenvalues are real and two pairs are complex conjugate. One complex conjugate eigenvalue pair $(-0.28191 \pm j8.6251)$ represents a local mode oscillation as discussed in chapter 2. In symbolic notation:

$$\lambda_{1,2} = \sigma \pm j\omega \quad (3.12)$$

The real part (σ) relates to damping and the imaginary part $(\pm\omega)$ relates to frequency of oscillation. Power engineers are more comfortable with damping ratio and frequency of oscillation in Hz. These are related to the eigenvalue by:

$$\zeta = -\frac{\sigma}{\sqrt{\sigma^2 + \omega^2}} \quad (3.13)$$

$$f = \frac{\omega}{2\pi} \quad (3.14)$$

The local mode is represented by a damping ratio of 0.03 and frequency of 1.37 Hz. Positive damping ratio indicates positive damping.

In a small system model with only a few hundred states this function can be used for quick calculation. In a large system with thousands of components only the important eigenvalues, those of the electromechanical modes, are computed and a vast calculation is avoided. This is known as selective modal analysis (SMA). One such technique is the AESOPS (Analysis of Essentially Spontaneous Oscillations in Power Systems) algorithm which was originally presented in [Byerley et al., 1982] and used in PEALS (Programme for Eigenvalue Analysis of Large Systems) [Kundur et al., 1990],[EL-5798,]. A problem with AESOPS is that the general characteristics of critical modes are not known but once they are known, it is easy to compute the eigenvalues and eigenvectors. It becomes an effective way of tracking the movement of critical eigenvalues as system conditions change. However, the modified Arnoldi method (MAM) overcomes this limitation [Wang and Semlyen, 1990] and it can compute the

Linear Control in Power Systems

eigenvalues of any system modes. Once the eigenvalues are computed, the corresponding eigenvectors are easily found.

3.2.1.1 Zero eigenvalues

Machine speeds and angles are expressed in absolute terms and redundancies occurs in the state variables and the state matrix is singular. If there is no infinite bus bar in the model there will be two zero eigenvalues. One arises from a redundancy in angle which may be removed by taking one machine angle as reference and expressing all other angles with respect to that. The angle state variable of the reference machine does not then appear in the system differential equations.

The second zero eigenvalue results when the generator torque is independent of machine speed deviations, that is, mechanical damping is neglected and governor action is not represented. This situation also arises when the ratio, inertia constant to damping coefficient, in all the machines is the same. This can be avoided by taking the speed of one machine as reference (assuming infinite inertia for this machine) and expressing the speed deviation of all the other machines with respect to it. The dynamics of the speed-reference machine no longer appears in the swing equation. In practice this is not done as it introduces difficulties in indexing and manipulating various matrices and vectors in vector-based computation. For a clear mathematical explanation, see [Sauer and Pai, 1998].

Practically, eigenvalues are not exactly zero as initial conditions of the states are not exact because of mismatches in power flow convergence, giving an approximate solution, albeit a good approximation.

3.2.1.2 Eigenvectors and modes

If λ_i is an eigenvalue and v_i is a non-zero vector such that the following relation holds:

$$Av_i = \lambda_i v_i \tag{3.15}$$

then the vector v_i is known as the right eigenvector of matrix A. In a matrix with all distinct eigenvalues (not a necessity but it is easier to understand when it is so), all the right eigenvectors and eigenvalues can be expressed as a compact matrix expression:

$$AV = V\Lambda \tag{3.16}$$

where,

$$V = \begin{pmatrix} v_1 & v_2 & \ldots & v_{n-1} & v_n \end{pmatrix} \tag{3.17}$$

$$\Lambda = diag\begin{pmatrix} \lambda_1 & \lambda_2 & \ldots & \lambda_{n-1} & \lambda_n \end{pmatrix} \tag{3.18}$$

Pre-multiplying both sides of (3.16) by V^{-1} gives

$$V^{-1}AV = \Lambda \qquad (3.19)$$

A similar expression holds for the left eigenvectors W such that

$$WA = \Lambda W \qquad (3.20)$$

where

$$W = \begin{bmatrix} w_1^t & w_2^t & .. & w_{n-1}^t & w_n^t \end{bmatrix}^t \qquad (3.21)$$

Post-multiplying both sides of (3.20) by W^{-1}, gives

$$WAW^{-1} = \Lambda \qquad (3.22)$$

The transformed physical state variables (x) can be put into modal variables (z) with the help of eigenvector matrices V and W

$$x = Vz \qquad (3.23)$$
$$z = Wx \qquad (3.24)$$

In power system literature, the right eigenvector matrix V is known as the mode shape matrix, that is, eigenvector v_i is known as the ith mode shape, corresponding to eigenvalue λ_i. The mode shape provides important information on the participation of an individual machine or a group of machines in one particular mode. Since in the example system we have considered an SMIB, the mode shape will be invariable and is of no interest. In a multi-machine system the mode shapes can identify coherent groups of machines.

3.2.1.3 Participation factors and eigenvalue sensitivities

In a large power system, it is important to quantify the role of each generator in each mode. It is natural to suggest that the significant state variables influencing a particular mode are those having large entries corresponding to the right eigenvector of λ_i. The problem of entries in an eigenvector is that they are impossible to compare because they have different units and scalings, that is, entries in the eigenvector corresponding to state variables such as speed, angle, flux, voltage etc. cannot be compared. Examining equations (3.16) and (3.20) or more precisely at the individual elements of the eigenvectors. v_{ki} relates the ith mode to the kth state variable, , that is, the activity of the ith mode in the kth state variable. On the other hand w_{ik}, relates the weighted contribution of the kth state variable in the ith mode. The product $w_{ki}v_{ik}$ is a dimensionless measure that is known as the participation factor [Verghese et al., 1982]. The left and right eigenvectors and their product $w_k v_i$ can be normalized to 1.0. If the same process is applied to obtain participation factors then they are known as

Table 3.1. Eigenvectors and normalized participation factors corresponding to local mode $-0.282 + j8.62$

left Eigenvector	right eigenvector	participation factor	participating state
$0.5492 - j\,4.0686$	$-0.0038 - j\,0.1149$	0.4372	machine angle
$0.4718 + j\,0.0009$	$+0.9923 - j\,0.0000$	0.4372	machine speed
$-1.9406 - j\,1.6424$	$-0.0165 - j\,-0.0297$	0.0800	field flux
$0.2853 - j\,2.5649$	$+0.0093 + j\,0.0047$	0.0248	q-axis damper
$-0.1986 + j\,0.1260$	$-0.0132 - j\,0.0013$	0.0029	d-axis damper
$-0.1159 + j\,1.3195$	$-0.0101 - j\,0.0054$	0.0140	q-axis damper
$0.4476 + j\,0.4505$	$-0.0102 + j\,0.0069$	0.0072	AVR
$-0.0012 + j\,0.0075$	$+0.0007 - j\,0.0137$	0.0001	SVC
$-0.0389 - j\,0.0069$	$-0.0015 - j\,0.0005$	0.0001	SVC
$0.0031 - j\,0.0068$	$-0.0045 + j\,0.0131$	0.0001	SVC

normalized participation factors. The more generic definition of participation factors for complex eigenvectors is obtained from [Sauer and Pai, 1998]:

$$p_{ki} = \frac{|v_{ik}|\,|w_{ki}|}{\sum_{k=1}^{k=n} |v_{ik}|\,|w_{ki}|} \quad (3.25)$$

Obviously the sum of such normalized participation factors is 1.0. The right, left eigenvector and the normalized participation factor is calculated for the local mode $(-0.282 \pm j8.62)$ and these are displayed in Table 3.1 for the example system shown in Fig. 3.1 The participation factor is an important tool in control design as it is used in identifying the participation of key generators in an inter-area mode. The generators with high participation factors are normally candidates for PSS. The state variable with the highest normalized participation factor can be the best choice for the feedback signal but it may not be a physical variable that can be measured. In Table 3.1, it is seen that machine angle and speed have very high normalized participation (0.4372) making them most effective as a feedback signal for control. However, machine angle cannot be measured physically but it can be synthesized from other measurements. On the other hand speed can be measured and has been used widely as a feedback signal in power system stabilizers (PSS).

It is also useful to see how an eigenvalue is sensitive to an individual element of the state matrix A. The basic eigenvalue and eigenvector relation is described as:

$$\lambda_i v_i = A v_i \quad (3.26)$$

Pre-multiplying both sides of (3.26) by the corresponding left eigenvector w_i gives

$$w_i \lambda_i v_i = w_i A v_i \qquad (3.27)$$

As, $w_i v_i = 1.0$

$$\lambda_i = w_i A v_i \qquad (3.28)$$

Differentiating with respect to the entry in the kth row and k^{th} column (a_{kk}) gives

$$\begin{aligned}
\frac{\partial \lambda_i}{\partial a_{kk}} &= \frac{\partial w_i}{\partial a_{kk}} A v_i + w_i \frac{\partial A}{\partial a_{kk}} v_i + w_i A \frac{\partial v_i}{\partial a_{kk}} \\
&= \frac{\partial w_i}{\partial a_{kk}} \lambda_i v_i + w_i \lambda_i \frac{\partial v_i}{\partial a_{kk}} + w_{ik} v_{ki} \\
&= \lambda_i \frac{\partial (w_i v_i)}{\partial a_{kk}} + w_{ik} v_{ki} \\
&= w_{ik} v_{ki}
\end{aligned} \qquad (3.29)$$

The final expression in (3.29) is obtained through the substitution of the relation $w_i v_i = 1.0$. It may be seen that the sensitivity of the eigenvalue to the diagonal element of a state matrix is a normalized participation factor as defined in (3.25). Eigenvalue sensitivity has been used to locate and design damping control in power systems. More on eigenvalue sensitivity can be found in [Pagola et al., 1989], [Rouco and Pagola, 1997], [Larsen et al., 1995], [Wang, 1999] and [Smed, 1993]

3.2.2 Modal controllability, observability and residue

One drawback of the participation factor approach is that it only deals with the states and does not include the input and output. It cannot effectively identify a controller site and an optimal feedback signal in the absence of information on the input and output which is more important when output feedback is employed. The effectiveness of control can however be indicated through controllability and observability indices. This is important as control cost is influenced to a great deal by the controllability and observability of the plant which can be addressed through modal controllability, observability and residue as described below. The transfer function equivalent of (3.7) and (3.9) is

$$G(s) = C(sI - A)^{-1} B + D \qquad (3.30)$$

Dropping the direct transmission term D which does not influence the mode (exclusion does not affect our conclusion but simplifies the explanation) and rewriting the first part of the right hand side (say $G_r(s)$ of (3.30) and making use of the orthogonal relationship between V and W, that is, $VW = I$

$$\begin{aligned} G_r(s) &= C(sI - A)^{-1}B \\ &= CVW(sI - A)^{-1}VWB \\ &= CV[V^{-1}(sI - A)W^{-1}]^{-1}WB \\ &= CV(sI - \Lambda)^{-1}WB \\ &= \sum_{i=1}^{i=n} \frac{Cv_i w_i B}{s - \lambda_i} \\ &= \sum_{i=1}^{i=n} \frac{R_i}{s - \lambda_i} \end{aligned} \qquad (3.31)$$

R_i is known as modal residue being the product of modal observability (Cv_i) and modal controllability ($w_i B$). It may be seen from (3.31) that modes with poor damping, that is, λ_i with small absolute real part, will significantly influence the magnitude of the transfer function G_r if it is scaled up by residue R_i. This means that the controllability of the input signal and the observability of the feedback signal are very important. The choice of feedback signal should broadly satisfy the following conditions:

3.2.2.1 Sensitivity to swing mode

The feedback signal must have a high degree of sensitivity at and around the swing mode frequency to be damped which will appear as a high peak in the Bode diagram. In other words, this means that the swing mode must be observable in the feedback signal.

3.2.2.2 Little or no sensitivity to other swing modes

This is an obvious aim from the perspective of minimizing the interaction between modes through the controller. A FACTS device in a transmission line will only influence those modes responsible for power swings observed on that line. Valuable control effort will be wasted if it responds to local swings within an area at one end of the line.

3.2.2.3 Little or no sensitivity to controller output

The feedback signal should have little or no sensitivity to its own output in the absence of a power swing. This is known as *inner loop sensitivity* [Larsen et al., 1995] and does not involve swing mode dynamics. It results from the feed-forward effect of a signal by-passing the swing mode loop. In a single-input-single-output (SISO) design, the output matrix C and input matrix B of equation (3.31) are row and column vectors respectively and hence the residue is a complex scalar. As the residue is a complex variable, both magnitude and phase are important. The higher the magnitude of the residue the less control efforts (gain) is needed and the higher the phase lag the more phase compensation blocks are needed in the feedback path.

For PSS siting, the method of residues can be used to identify the best location, that is, the candidate machine where the PSS is most effective. One important aspect to be kept in mind is that any choice based on the comparison of feedback signals must be restricted to the same type of signal. Even though machine currents, speed and power are expressed in p.u., residues, as computed based on each of these signals as an output candidate, they must not be compared against one another. Residue obtained for a particular type of signal (say power) must not be compared with that obtained for other signal which is not power. The generator whose output power produces the largest residue magnitude would be the best PSS location requiring the least control effort. Having said this, the residue obtained from machine power as feedback signal can be compared with that obtained from speed as feedback signal if the residue based on speed is scaled by a factor $(D + M\lambda_i)$ where D and M are the mechanical damping and inertia constants of the generator. This simple relation follows from the swing equation with constant mechanical input.

In damping control design employing FACTS devices, the method of residues plays an important role. Usually FACTS devices are placed in the network from steady-state operational consideration such as power flow (series devices), dynamic voltage control (shunt devices). The locations are decided based on studies considering a large number of power flow scenarios including network topologies. This fixes the modal controllability but it is necessary to compute the modal observability of power flows in different lines originating from or terminating at the FACTS location. Remote signals or a combination of synthesized signals are also possible options but in these cases it becomes a centralized design.

In the model system, we have chosen generator speed, generator output power and line power between bus 2 and 3 as candidate feedback signals for a SVC damping controller. The voltage reference to the SVC is taken as input and the computed modal controllability, observability and residues are shown in Table 3.2. It can be seen that as the SVC location is fixed at bus 2, the modal controllability is also fixed, but the modal observability varies with the signals. It is obvious that generator output power or the active power in the line between bus 2 and 3 is the most effective in damping the local mode when compared with generator speed after proper scaling. In this case both of them are equally effective as the whole of the generator real power output flows through the line between bus 2 and 3 in the absence of any load at the SVC bus location. As

Linear Control in Power Systems

Table 3.2. Modal controllability, observability and residue corresponding to local mode $-0.282 + j8.62$

feedback signal	controllability	observability	residue
generator speed	$0.3798\angle - 105^0$	$0.9923\angle 0^0$	$0.3769\angle - 105^0$
generator power	$0.3798\angle - 105^0$	$3.5590\angle - 97.6^0$	$1.3517\angle 157.55^0$
line power (P_{23})	$0.3798\angle - 105^0$	$3.5590\angle - 97.6^0$	$1.3517\angle 157.55^0$

the generator output power is a remote signal, it would be more reliable to have the local line power. In subsequent chapters, we will discuss the use of remote signals in more detail when dealing with FACTS control design and analysis through the techniques of multivariable control.

For a large practical power system model, sparsity techniques are exploited in the differential and algebraic structure of the system described in (3.4) and (3.5) for fast computation of modal controllability, observability and residue [Martins and Lima, 1990].

3.2.3 Singular values and singular vectors

Although eigenvalues are a good guide to understanding the modal behavior of a system, their magnitudes do not provide useful means of generalizing gain $|G(j\omega)|$ of a SISO system [Skogestad and Postlethwaite, 2001]. Eigenvalues can only be computed for a square matrix and even then they can be very misleading. To clarify this, let us consider a system $y = Gd$ with

$$G = \begin{bmatrix} 0 & 0 \\ 50 & 0 \end{bmatrix} \quad (3.32)$$

The eigenvalues are both zeros. This does not mean that the system gain is zero as one can have output $y = [0 \; 50]^T$ with input $d = [1 \; 0]^T$. The eigenvalues are indicative of system gains when the input and the output are in the same direction and in the direction of eigenvectors. Let us assume G has eigenvalue λ and eigenvector t_i which is also chosen as input d. The output is $y = Gt_i = \lambda_i t_i$. We get $\|y\|/\|d\| = \|\lambda_i t_i\|/\|t_i\| = |\lambda_i|$, so $|\lambda_i|$ measures gain in the direction t_i. This is not useful in performance analysis although it could be quite handy in stability assessment.

The singular values, on the other hand, provide gain in the principal directions. Let us define singular value decomposition (SVD): Any $l \times m$ matrix G can be expressed as

$$G = U \sum V^H \quad (3.33)$$

where $\sum$ is an $l \times m$ matrix with $k = min(l, m)$ non-negative singular values, σ_i, arranged in descending order along its main diagonal; the other entries are zeros. U is an $l \times l$ matrix of output singular vectors u_i and V is an $m \times m$ unitary matrix of input singular vectors v_i. V^H is the complex conjugate transpose of V. Usually we use U as input and V as voltage or right eigenvector in this book. However, these are also standard notations for singular vectors but as the context is very clear no confusion should arise. The singular values are positive square roots of the eigenvalues of $G^H G$ which are real because the matrix $G^H G$ is symmetric, or in other words

$$\sigma_i(G) = \sqrt{\lambda_i(G^H G)} \tag{3.34}$$

Each column of u_i of output matrix U are orthonormal (unit length), that is

$$\|u_i\|_2 = \sqrt{|u_{i1}|^2 + |u_{i2}|^2 + \ldots\ldots + |u_{il}|^2} = 1$$
$$u_i^H u_i = 1 \quad u_i^H u_j = 0, i \neq j \tag{3.35}$$

u_i indicates the ith principal *output direction*. Similarly each column v_i in V is orthonormal and represents the ith *input direction*. These input and output directions are connected through the singular value σ_i. To see this, equation (3.34) can be written and simplified, using relation $V^H V = I$ to give

$$G v_i = \sigma_i u_i \tag{3.36}$$

This suggests that if an input is considered in the direction of v_i the output is in the direction u_i. Since $\|u_i\|_2 = 1$ and $\|v_i\|_2 = 1$, it is seen that σ_i is the gain of the transfer matrix in this direction. In essence,

$$\sigma_i(G) = \|G v_i\|_2 = \frac{\|G v_i\|_2}{\|v_i\|_2} \tag{3.37}$$

Besides giving better information about the gains of the plant and being applicable to non-square plants, the directions given by the singular vectors are orthogonal. It is worth expanding the relation in equation (3.37) when the singular value is σ_1 that is the first singular value from the diagonal matrix $\sum$ in equation (3.33). This, by definition, is the largest one and hence is termed the maximum singular value, given by

$$\bar{\sigma}(G) \equiv \sigma_1(G) = \max_{d \neq 0} \frac{\|G d\|_2}{\|d\|_2} = \frac{\|G v_1\|_2}{\|v_1\|_2} \tag{3.38}$$

Similarly the lowest gain in any input direction is the minimum singular value

$$\underline{\sigma}(G) \equiv \sigma_k(G) = \max_{d \neq 0} \frac{\|G d\|_2}{\|d\|_2} = \frac{\|G v_k\|_2}{\|v_k\|_2} \tag{3.39}$$

Linear Control in Power Systems

It is needless to mention that the most effective input direction is v_1 and the most effective output direction is u_1. This is very useful from the perspective of multi-variable control where often the cost of control becomes a constraining factor in overall control design.

The ratio of maximum singular value to minimum singular value is known as condition number ($\gamma(G)$) of the plant.

$$\gamma(G) = \frac{\overline{\sigma}(G)}{\underline{\sigma}(G)} \tag{3.40}$$

The condition number is an important frequency domain property of a plant and gives an indication of the degree of difficulty to control it. Usually a large condition number means the plant is ill-conditioned, that is, harder to control. However, a large condition number does not mean [Skogestad and Postlethwaite, 2001] that the plant is very sensitive to uncertainty but the reverse can be true, that is, if a plant has a small condition number then the multi-variable effects of uncertainty are not so serious.

3.2.4 $\mathcal{H}_\infty$ and $\mathcal{H}_2$ norm

The $\mathcal{H}_\infty$ norm of a system is the peak value of the magnitude of the transfer function over the whole frequency range. In a SISO system, it is the peak value in Bode magnitude plot. In a MIMO system, it is taken as the peak value of the maximum singular value response as function of frequency and expressed as:

$$\|G(s)\|_\infty \stackrel{\Delta}{=} \max_\omega \overline{\sigma}(G(j\omega)) \tag{3.41}$$

Since the singular value provides maximum gain in the principal direction, $\mathcal{H}_\infty$ norm is seen as the magnitude of some loop transfer function in the worst direction over the entire frequency range.

The $\mathcal{H}_\infty$ norm also admits some nice properties in the time domain. It is equal to the induced norm in the time domain; that is, given input signal $w(t)$ and output signal $z(t)$, the $\mathcal{H}_\infty$ norm indicates the ratio of the Euclidian norm of these two signals for any time

$$\|G(s)\|_\infty \stackrel{\Delta}{=} \max_{w(t) \neq 0} \frac{\|z(t)\|_2}{\|w(t)\|_2} \tag{3.42}$$

The $\mathcal{H}_\infty$ norm is also the induced power norm in terms of the expected values of stochastic signals. All these nice properties make the $\mathcal{H}_\infty$ norm very useful in various engineering applications. The plot of maximum singular value gain response shown in Fig 3.2 for the example system shows the value of $\mathcal{H}_\infty$ norm to be 2.3407.

The $\mathcal{H}_2$ norm is a measure of overall energy of a system relating input disturbance to output response. The formal definition in the frequency domain is [Skogestad and Postlethwaite, 2001]:

$$\|G(s)\|_2 \triangleq \sqrt{\frac{1}{2\pi}\left(\int_{-\infty}^{+\infty} \text{tr}\underbrace{\left(G(j\omega)^H G(j\omega)\right)}_{\|G(j\omega)\|_F^2 = \sum_{ij}|G_{ij}(j\omega)|^2} d\omega\right)} \quad (3.43)$$

where, $tr(A)$ is the sum of the diagonal of A. It is seen from the definition that the norm is of finite value for a strictly proper system. The $\mathcal{H}_2$ norm has also another interpretation as the $\mathcal{H}_2$ norm of the impulse response $g(t)$ of $g(s)$ and is given by

$$\|G(s)\|_2 = \|g(t)\|_2 \triangleq \sqrt{\left(\int_0^{+\infty} \text{tr}\underbrace{\left(g^T(\tau)g(\tau)\right)}_{\|g(\tau)\|_F^2 = \sum_{ij}|g_{ij}(\tau)|^2} d\tau\right)} \quad (3.44)$$

The equation (3.44) can be reorganized as:

$$\|G(s)\|_2 = \|g(t)\|_2 \triangleq \sqrt{\left(\sum_{ij}\int_0^{+\infty}|g_{ij}(\tau)|^2 d\tau\right)} \quad (3.45)$$

where $g_{ij}(t)$ is the ij^{th} element of the impulse response matrix, $g(t)$. Hence the $\mathcal{H}_2$ norm can be interpreted as the output resulting from applying an unit impulse to each input one after another, that is, allowing the output settle to zero before applying next input and so on. In summary, the deterministic performance interpretation of the $\mathcal{H}_2$ norm is given by the following expression:

$$\|G(s)\|_2 = \max_{\omega(t)=\text{unitimpulse}} \|z(t)\|_2 \quad (3.46)$$

Numerically $\mathcal{H}_2$ norm can be obtained from one of the relations $\|G(s)\|_2 = \sqrt{\text{tr}(B^T Q B)}$ or $\|G(s)\|_2 = \sqrt{\text{tr}(CPC^T)}$. The P and Q are the observability and controllability grammian respectively that are obtained from the solution of appropriate Lyapunov equations. The $\mathcal{H}_2$ norm in this example system is obtained as 1.3261. It is seen that it is less than the $\mathcal{H}_\infty$ norm.

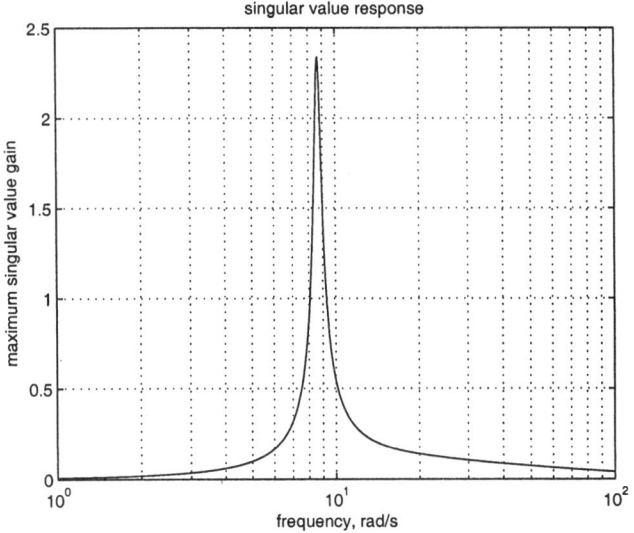

Figure 3.2. maximum singular value response of the example system

3.2.5 Hankel singular values and model reduction

Consider a state space realization of a system $G(s) \stackrel{s}{=} (A, B, C, D)$. The solution P and Q of the system of equations

$$PA^T + AP + BB^T = 0 \qquad (3.47)$$
$$QA + A^TQ + C^TC = 0 \qquad (3.48)$$

are known as the controllability and observability grammian respectively. The singular value of the product of controllability and observability grammian is $\sigma(PQ) = \text{diag}(\sigma_1, \sigma_2 ... \sigma_n)$. The σ are Hankel singular values (HSV) of $G(s)$ and ordered as $\sigma_1 > \sigma_2 > \sigma_3 ... \sigma_n$. In other words $\sigma_i = \lambda_i(PQ)^{1/2}$. The largest singular value σ_1 by definition is the Hankel norm of the system, that is, $\|G\|_H = \sigma_1$. In the example system, taking the line power as output and the SVC voltage reference as input, the HSV can be found as:

$$HSV = \begin{bmatrix} 1.215 & 1.18 & 0.06 & 0.026 & 0.002 & 0.002 & 0.0003 & 0.0001 & 0 & 0 \end{bmatrix}^T \qquad (3.49)$$

The order of the controllers, that is, the number of state variables, synthesized using $\mathcal{H}_\infty$ norm minimization techniques is at least as high as the order of the open-loop system although it might be even higher with the incorporation of weighting functions. Therefore, it is mandatory to simplify the system model first, if possible, to ease the design procedure and to avoid complexity in the final controller. The simplified system used for design must, however, be a good approximation of the original system. This is known as the 'model order reduction problem' where the central idea is 'given a high-order (say n) system model $G(s)$, derive a low order approximation (say r) $G_r(s)$ such that the infinity norm of their difference $\|G(j\omega) - G_r(j\omega)\|_\infty$ is sufficiently small. The same idea is also applicable for simplifying the controller once it is designed. A number of methods for model reduction are available in control literature and some are described below.

3.2.5.1 Modal truncation

This is a simple approach where the less significant part of the model is ignored. Let us look at (3.31). If the eigenvalues $\lambda_i s$ are ordered such that $|\lambda_1| < |\lambda_2|...|\lambda_r| < |\lambda_{r+1}|... < |\lambda_n|$ and the term $\overline{\sigma}\left(\sum_{i=r+1}^{i=n} \frac{R_i}{s-\lambda_i}\right)$ is very small, then the model $G(s)$ can be approximated by the first r eigenvalues of the system. The error depends on the residue terms and the real part of the corresponding eigenvalues $\lambda_{r+1}...\lambda_n$. The error in the high frequency range is small as the truncated model $G_r(\infty) = G(\infty)$ at infinite frequency and equals the dc gain D. It is possible to reduce the model when the matrix is transformed into a modal co-ordinate, that is, in Jordan canonical form.

3.2.5.2 Residualization

The state space realization (A, B, C, D) is partitioned into a group of slower (smaller λ) and faster states (larger λ) which can be written as follows:

$$\dot{x}_1 = A_{11}x_1 + A_{12}x_2 + B_1 u \quad (3.50)$$
$$\dot{x}_2 = A_{21}x_1 + A_{22}x_2 + B_2 u \quad (3.51)$$
$$y = C_1 x_1 + C_2 x_2 + D u \quad (3.52)$$

Unlike in modal truncation, the left side of (3.51) is set to zero and the variables x_2 is eliminated from (3.50) and (3.52). The reduced order model $G_r(s) \stackrel{s}{=} (A_r, B_r, C_r, D_r)$ is called the *balanced residualization* with

$$A_r = A_{11} - A_{12} A_{22}^{-1} A_{21}$$
$$B_r = B_1 - A_{12} A_{22}^{-1} B_2$$
$$C_r = C_1 - C_2 A_{22}^{-1} A_{21}$$
$$D_r = D - C_2 A_{22}^{-1} B_2$$

$$(3.53)$$

Linear Control in Power Systems

The error in this model reduction is zero at DC and maximum at high frequencies because the faster dynamics are ignored. This is also known as singular perturbation. In power system modelling this method is exclusively applied ignoring stator and network transients, giving rise to the standard DAE form. The vector of algebraic variables y is equivalent to x_2 in this formulation [Chow, 1983].

3.2.5.3 Balanced truncation and residualization

The success of both modal truncation and residualization depends largely on the original realization and it is most convenient when expressed in Jordan form. However, *balanced realization* is a technique that comes with many useful properties. A realization $G(s) \stackrel{s}{=} (A, B, C, D)$ is said to be balanced when the controllability grammian P and observability grammian Q defined in (3.47) and (3.48) are equal and diagonal. The entries in the diagonal are the HSV of $G(s)$. Any minimal realization of a transfer function can be balanced through a series of similarity transformations [Laub et al., 1987]. The most useful aspect of minimal realization is that it relates each transformed state to a HSV. The value of the HSV indicates to what extend that equivalent state influences the frequency response of $G(s)$. If $\sigma_r \gg \sigma_{r+1}$, the effect of state x_r in input-output behavior is greater compared to that of x_{r+k}. In balanced realization, each state is as equally controllable as it is observable. The HSV, being the product of the two grammians, can be treated as an index that provides an idea as to what extent a system model can be reduced a priori. Accordingly, the balanced realization is partitioned into a group of the most significant states and the least significant states with poor controllability and observability, respectively. The modal truncation is then applied to obtain the reduced model by ignoring the least significant states. This is known as *balanced truncation* [Moore, 1981]. The DC gain is preserved and accuracy is better at the high frequency end. The dynamics of the least significant group of states are set to zero and treated as algebraic variables to get the reduced model as described earlier. This is known as *balanced residualization* [Fernando and Nicholson, 1982] and in this method, the error is least in the low frequency range. Hence, depending on the frequency range of interest, one can be better than the other. Mathematically, one can be transformed to the other through a simple bilinear transformation $s \to \frac{1}{s}$. In either of these methods, the guaranteed bound on error of approximation is expressed as twice the sum of the HSV ignored (known as twice the sum of the tail), that is,

$$\|G(j\omega) - G_r(j\omega)\|_\infty \leq 2 \sum_{i=k+1}^{n} \sigma_i \qquad (3.54)$$

A closer look at the HSV in equation (3.49) reveals that only the first 2 out of 10 HSV are significant. This gives an indication that the 10th order example system can be reduced to a 2nd or 3rd order one.

However, sometimes, a balancing transformation matrix can be ill-conditioned. This is best addressed via Schur's method of decomposition of a grammian product PQ for computing an orthogonal transformation operator which is numerically robust [Safonov and Chiang, 1989]. In this book, the reduction of the power system model is carried out using this method implemented in the *'schmr'* function available in *Robust control toolbox* [Packard et al., 2002] in *Matlab* [mat, 1998].

3.2.5.4 Optimal Hankel norm approximation

In this approach, the order r of the reduced model is obtained by solving an optimization problem [Glover, 1984]: Given a model $G(s)$ of order n, find a reduced order model of degree k such that the Hankel norm of the error $\|G(s) - G_k(s)\|_H$ is minimized. In other words, the optimization routine seeks to minimize the largest HSV of the error transfer functions.

We have applied four model reduction techniques as described above to reduce the model of the example system. The order of the reduced model is chosen to be four and the frequency response of the reduced model is displayed in Fig 3.3. It is seen that in all four methods, the errors in the reduced model are almost similar except in balanced residualization which shows higher accuracy in the low frequency region. When the order of the reduced system is chosen as three, the errors increase which can be observed in Fig 3.4.

This exercise shows that the HSV gives a good indication of the degree to which system model complexities can be simplified.

However, when the size of the system becomes larger (say an order of more than 1000), solving the Lyapunov equations to compute P and Q becomes difficult. In those cases, approximate solutions obtained by the methods of Krylov subspace have been found very useful [Jaimoukha and Kasenally, 1997, Chaniotis and Pai, 2005]. In current research, we found that for a system with about 380 states, the Krylov subspace based produced results much faster (about 85 times) than balanced truncation, but suffered from a larger error. This is shown in terms of the frequency response of the original and reduced system in Fig. 3.5.

Our strategy has been to use Krylov subspace to reduce the order to less than 100 and then apply Schur's balanced truncation approach for better speed and accuracy combined as shown in Fig. 3.6.

Linear Control in Power Systems

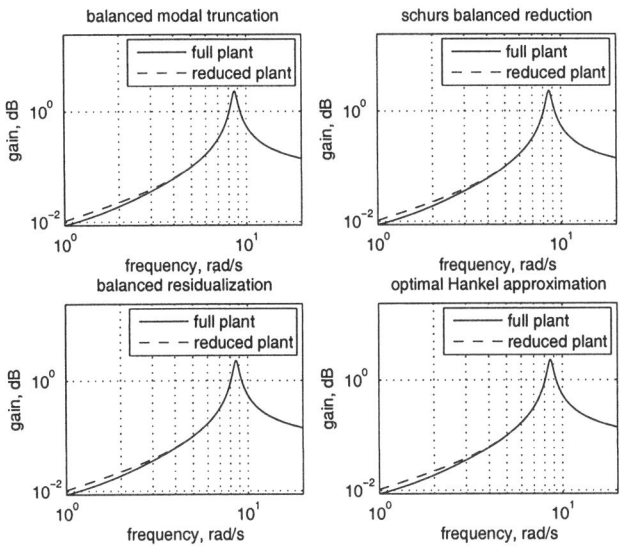

Figure 3.3. Frequency response of the original and the reduced 4th order system

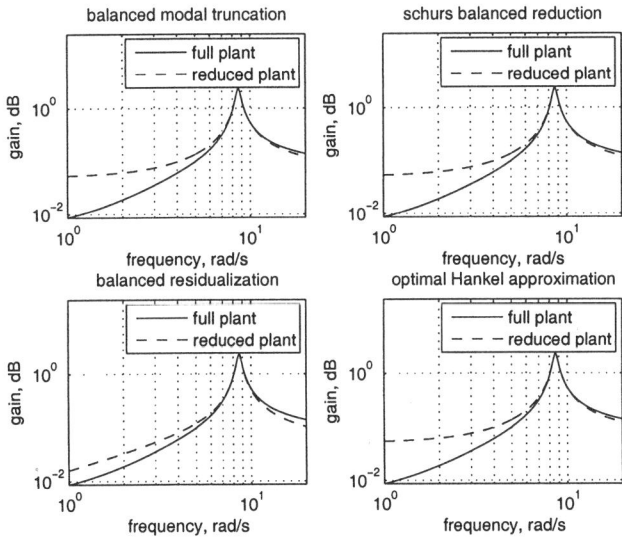

Figure 3.4. Frequency response of the original and the reduced 3rd order system

3.2.6 Stability, performance and robustness

There have been many formal definitions of stability in the control literature. However, what we are concerned about here is the bounded behavior of the

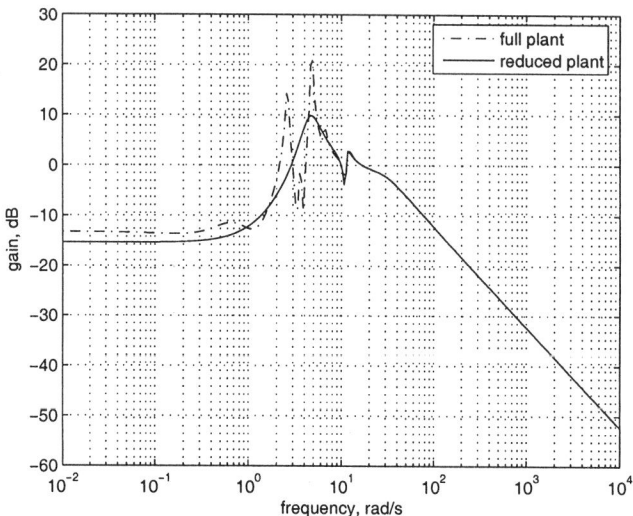

Figure 3.5. Frequency response of the original (365 states) and reduced (20 states) system using Krylov subspace based techniques

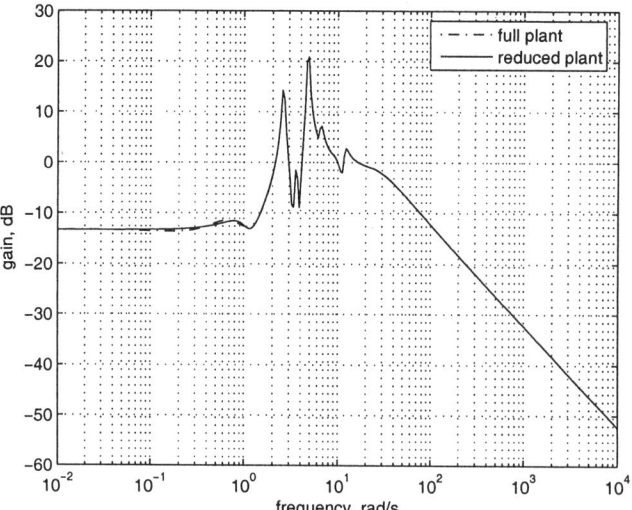

Figure 3.6. Frequency response of the original (365 states) and reduced (20 states) system: simplified down to 100 states using Krylov subspace based techniques and then to 20 states using balanced truncation

system when subjected to bounded disturbances. It is not only input-output stability, but rather internal stability [Skogestad and Postlethwaite, 2001]that is of importance. Disturbances in all possible input channels or input points available internally between all the components of the system are applied and the response is all components including the accessible output channels are observed. If the response in the whole set is bounded, the system is said to be internally stable but if all the hidden modes (non controllable and non observable) are stable then the system is also state stabilizable and state detectable, otherwise not.

In the context of power system stability the most recently proposed definition is [Kundur et al., 2004] *"Power system stability is the ability of an electric power system, for a given initial operating condition, to regain a state of operating equilibrium after being subjected to a physical disturbance, with most system variables bounded so that practically the entire system remains intact"*. Power system stability phenomena has been classified in different time frames for observation, based on the severity of the disturbances and the state variables , that is, angle,voltage and frequency. Accordingly, large disturbance angle stability and small disturbance angle stability are defined. Transient stability, also known as rotor angle swing in the early days, is a short term phenomenon related to that of maintaining synchronism. In some situations, synchronism is not a problem but the voltage gradually collapses when loading reaches a very high value. This type of rundown situation is classified as voltage stability which is also classified as large disturbance, small disturbance or short/long term phenomenon. Frequency stability is also categorized as both short term and long term phenomena. Generally, angle instability is related to runaway situations and voltage instability (collapse) is related to rundown situations.

Performance in a control system incorporates functions which make the output behave in a more desirable manner. It can be specified in both time and frequency domain. The usual practice is to test the speed and the quality of the response of the closed-loop system for a step disturbance. *Rise time, settling time, percentage overshoot, steady state offset* etc. are the measures of performance. In the frequency domain, loop transfer function or various closed loop transfer functions characterize closed-loop performance. The main advantage of the frequency domain approach compared to step response is that it relates to a broader class of signals (sinusoids of any frequency). These properties are termed as disturbance rejection, good tracking and noise attenuation etc. thereby making characterization of the feedback property easier. Some of the frequency domain measures to assess the margin of stability of a system are gain and phase margin, the maximum singular values of suitably weighted sensitivity (S) and complementary sensitivity (T) matrices in a multivariable system. These symbols are standard and will be used extensively in the later chapters. The speed of the response in the frequency domain is assessed through crossover and bandwidth frequency. The higher the crossover frequency, the greater is the bandwidth. Larger bandwidth gives rise to better performance and the response of the system is smaller. Usually the product of bandwidth and response time is fairly constant. However, greater bandwidth gives rise to larger transient overshoot, implying less stability gain and phase margin or more closer operation to the critical point in a Nyquist diagram.

A control system is robust if it has no or very little sensitivity to the difference between the actual system in practice and its model used for control design. These are termed as model mismatch or uncertainties. The key idea in robust control is to maximize this uncertainty over which the performance of the control is deemed adequate. The uncertainty modelling in robust control literature is all about finding a frequency domain mathematical expression for this mismatch. Uncertainties can rarely be expressed as an explicit factor on top of the nominal plant model. Usually they are nonuniform across all the channels and are expressed as either structured or unstructured perturbations around the plant model in the frequency domain. The dynamic behavior of a power system during a disturbance is difficult to model in the frequency domain as there is no model of uncertainty connecting pre-fault to post-fault steady state. The robustness margin of a controller or closed loop system is designed by including these perturbations via weighting functions and a quantitative measure expressed as $\mathcal{H}_\infty$ norm computed for which the closed-loop system provides the desired behavior. Robustness is applicable to both stability and performance of a system. By robust stability (RS) is meant that whether a system remains stable over a set of all plants in the uncertainty set. Robust performance (RP), means that performance specifications are met for all plants in the uncertainty set besides satisfying RS. In the context of power system damping control design, robustness means whether the damping is adequate and the oscillations decay quickly when time domain simulations are performed for all likely disturbances under all significant operating conditions.

3.2.7 Control design specifications in power systems

It is difficult to translate power system operating objectives into standard control design specifications such as tracking, disturbance rejection, noise attenuation etc. However, a power system damping controller must be designed to address the following performance, stability and robustness criteria

- A minimum damping ratio must be maintained for the critical modes. Different utilities set different minimum values. Ontario Hydro practice [Kundur, 1994] is to have a damping ratio of 0.03 whereas Australian utilities use

0.05. However,low frequency mode(such as interarea modes) require better damping ratios (greater than 0.1) [Pal, 1999].

- The oscillations must settle within a desired time as set out in the utility operating guidelines. The utilities in a few countries adopt settling time of oscillations as damping criterion,that is, a specification in time domain. The settling time varies from one utility to other. In the UK and New Zealand, a settling time of 10-12 seconds is specified. In the Danish and Norwegian systems, this value ranges between 10-20 seconds. More detail on this can be found in [Pal, 1999]

- The damping must not deteriorate to unacceptable levels at different operating conditions and network configurations. This is a robustness measure of performance and a measure of stability margin, as with lower damping ratio the phase margin is reduced. Interconnections are now required not only to meet the criteria appropriate for maximum rated power transfer but to meet all the conditions that arise including those arising from trading arrangements.

- The controllers for different devices in the system should not interact adversely. Until recent years, the industry did not force a strict requirement on coordination among power system damping controllers. Now, in the light interconnection causing blackouts, the focus is more towards coordinated control of the system. Coordinated design of PSS and PSS and FACTS have received attention in the recent decades. The multivariable approach to control design can address this issue much better than the classical approach through sequential design.

There are many design methods described in the literature but the characteristics of a power system through a DAE dynamical description imposes a number of difficulties in employing them to large practical system models. According to CIGRE task force on *"Impact of Interactions amongst power system controls"* [Martins, 2000] the requirement on control design methods for multiple controllers in power systems can broadly be summarized into four categories,as follows.

3.2.7.1 Coordinated design

Coordination does not mean centralized design, but rather simultaneous tuning of parameters of a number of decentralized controllers to ensure dynamic and steady state performance criteria are met whilst minimizing or preventing deleterious interactions among controllers.

3.2.7.2 Practical control system structures

The complexity of control structures can be a limiting factor to the usefulness of the control design method. For example, industry prefers decentralized structure with output feedback. However, it is likely they will adopt other control structures to eliminate the limitations of output-feedback based decentralized structures.

3.2.7.3 Emerging techniques for control design

Power system dynamic behavior in the low frequency range is described through a set of sparse DAE. The DAE form has the advantage that sparsity can be exploited. It is therefore, very useful to have a control design that builds around this structure. This is a challenging problem in control theory. If control techniques keep to this form in future, they can be applied to large practical power system models. The other approach is a reduction in the form of elimination of algebraic variables at the loss of sparsity. All the techniques of robust control applied to power system damping design that we have come across are in this category. The trouble is that a controller with a large order results but its performance is demonstrated to be better.

3.2.7.4 Robustness

The control system must provide adequate damping and security margin in all operating conditions and network configurations that may be required.

3.3 Summary

We have provided an overview of the various analytical tools and techniques of linear system theory that is used in power system control. These have been demonstrated in an example power system where these tools are applied. This was an attempt to provide a better understanding of these tools from the view point of power system engineers. Recently defined terms and definitions of stability, performance and robustness have been described. The requirements and specifications in power system control design have also been discussed.

References

[mat, 1998] (1998). *Matlab Users Guide*. The Math Works Inc., USA.

[Byerley et al., 1982] Byerley, R.T., Bennon, R.J., and Sherman, D.E. (1982). Eigenvalue analysis of synchronising power flow oscillations in large electric power systems. *IEEE Transactions on Power Apparatus and Systems*, PAS-101:235–243.

REFERENCES

[Chaniotis and Pai, 2005] Chaniotis, D. and Pai, M.A. (2005). Model reduction in power systems using krylov subspace methods. *IEEE Transactions on Power Systems*. accepted for publication.

[Chow, 1983] Chow, J.H. (1983). *Time-scale Modeling of Dynamic Networks with Applications to Power Systems*. Springer-Verlag Publishers, New York.

[EL-5798,] EL-5798, EPRI Report. *The Small Signal Stability Programme Package*. EPRI.

[Fernando and Nicholson, 1982] Fernando, K.V. and Nicholson, H. (1982). Singular perturbational model reduction for balanced systems. *IEEE Transactions on Automatic Control*, 27(2):466–468.

[Fouad and Vittal, 1992] Fouad, A.A. and Vittal, V. (1992). *Power System Transient Stability Analysis Using the Transient Energy Function Method*. Prentice-Hall, USA.

[Glover, 1984] Glover, K. (1984). All optimal hankel norm approximations of linear multivariable systems and their l^∞ error bounds. *International journal of control*, 39(6):1115–1193.

[Jaimoukha and Kasenally, 1997] Jaimoukha, I.M. and Kasenally, E.M. (1997). Implicitly restarted krylov subspace methods for stable partial realizations. *SIAM J. Matrix Analysis and Applications*, 18(3):633–652.

[Kundur, 1994] Kundur, P. (1994). *Power System Stability and Control*. McGraw Hill, USA.

[Kundur et al., 2004] Kundur, P., Paserba, J., Ajjarapu, V., Andersson, G., Bose, A., Canizares, C., Hatziargyriou, N., Hill, D., Stankovic, A., Taylor, C., and Vittal, T. Van Cutsemand V. (2004). Definition and classification of power system stability ieee/cigre joint task force on stability terms and definitions. *IEEE Transactions on Power Systems*, 19(3):1387–1401.

[Kundur et al., 1990] Kundur, P, Rogers, G.J, Wong, D.Y., Wang, L., and Lauby, M.G. (1990). A comprehensive computer program for small signal stability analysis of power systems. *IEEE Transactions on Power Systems*, 5(4):1076–1083.

[Larsen et al., 1995] Larsen, E.V., Sanchez-Gasca, J.J., and Chow, J.H. (1995). Concepts for design of facts controllers to damp power swings. *Power Systems, IEEE Transactions on*, 10(2):948–956.

[Laub ct al., 1987] Laub, A.J., Heath, M.T., Page, C.C., and Ward, R.C. (1987). Computation of balancing transformations and other applications of simultaneous diagoninalization algorithms. *IEEE Transactions on Automatic Control*, 32(2):115–122.

[Martins, 2000] Martins, N. (2000). Impact of interactions among power system controls. *CIGRE Special Publication 38.02.16*, Technical Brochure 166.

[Martins and Lima, 1990] Martins, N. and Lima, L.T.G. (1990). Determination of suitable locations for power system stabilizers and static var compensators for damping electromechanical oscillations in large power systems. *IEEE Transactions on Power Systems*, 5(4):1455–1469.

[Moore, 1981] Moore, B.C. (1981). Principal component analysis in linear systems: controllability, observability and model reduction. *IEEE Transactions on Automatic Control*, 26(1):17–32.

[Packard et al., 2002] Packard, A., Balas, G.J., Safonov, M., and R., Chiang (2002). *Robust Control Toolbox for use with Matlab*. The Math Works Inc., USA.

[Pagola et al., 1989] Pagola, F.L., Perez-Arriaga, I.J., and Verghese, G.C. (1989). On sensitivities, residues and participations: applications to oscillatory stability analysis and control. *Power Systems, IEEE Transactions on*, 4(1):278–285.

[Pai, 1989] Pai, M.A. (1989). *Energy Function Analysis for Power System Stability*. Kluwer Academic Publishers, USA.

[Pal, 1999] Pal, B.C. (1999). *Robust Damping Control of Inter-area Oscillations in Power System with Super-conducting Magnetic Energy Storage Devices*. PhD thesis, Imperial College of Science Technology and Medicine, Department of Electrical and Electronic Engineering.

[Pavella and Murthy, 1994] Pavella, M. and Murthy, P.G. (1994). *Transient Stability of Power Systems: Theory and Practice*. John Willey and Sons, Chichester.

[Rouco and Pagola, 1997] Rouco, L. and Pagola, F.L. (1997). An eigenvalue sensitivity approach to location and controller design of controllable series capacitors for damping power system oscillations. *IEEE Transactions on Power Systems*, 12(4):1660–1666.

[Safonov and Chiang, 1989] Safonov, M.G. and Chiang, R.Y. (1989). A schur method for balanced-truncation model reduction. *Automatic Control, IEEE Transactions on*, 34(7):729–733.

[Sauer and Pai, 1998] Sauer, P.W. and Pai, M.A. (1998). *Power System Dynamics and Stability*. Prentice Hall, USA.

[Skogestad and Postlethwaite, 2001] Skogestad, S. and Postlethwaite, I. (2001). *Multivariable Feedback Control*. John Wiley and Sons, UK.

[Smed, 1993] Smed, T. (1993). Feasible eigenvalue sensitivity for large power systems. *IEEE Transactions on Power Systems*, 8(2):555–563.

[Verghese et al., 1982] Verghese, G.C., Perez-Arriaga, I.J., and Scheweppe, F.C. (1982). Selective modal analysis with applications to electric power systems, part i and part ii. *IEEE Transactions on Power Apparatus and Systems*, PAS-101:3117–3134.

[Wang, 1999] Wang, H.F. (1999). Selection of robust installing locations and feedback signals of FACTS-based stabilizers in multi -machine power systems. *IEEE Transactions on Power Systems*, 14(2):569–574.

[Wang and Semlyen, 1990] Wang, L. and Semlyen, A. (1990). Application of sparse eigenvalues techniques to the small signal stability analysis of large power systems. *IEEE Transactions on Power Systems*, 5(2):635–642.

Chapter 4

TEST SYSTEM MODEL

In power systems, the primary sources of electrical energy are the synchronous generators. The problem of power system stability is primarily to keep the interconnected synchronous machines in synchronism [Kundur, 1994]. The stability is also dependent on several other components such as the speed governors, excitation systems of the generators, the loads, the FACTS devices etc.. Therefore, an understanding of their characteristics and modelling of their performance are of fundamental importance for stability studies and control design. The general approach to modelling of several power system components is quite standard, and a quick overview of these models is given in this chapter with respect to a particular study system.

A 16-machine, 68 bus test system is considered for the illustration of different control design techniques. This chapter presents an overview of the test system including the models used to describe various components (e.g. generators). The numerical data for different model parameters is provided in the Appendices A and B at the end of the book.

The dynamic behavior of the power system is described by a set of non-linear differential-algebraic equations (DAE). Linearization of these DAEs about an operating point to obtain the linearized system matrix is described in Section 4.4. The critical eigen values of the system are also shown. Appropriate selection of the feedback signals corresponding to the critical modes of the system is discussed in Section 4.5 and simplification of the linearized system model, which is a pre-requisite to control design, is described in Section 4.6.

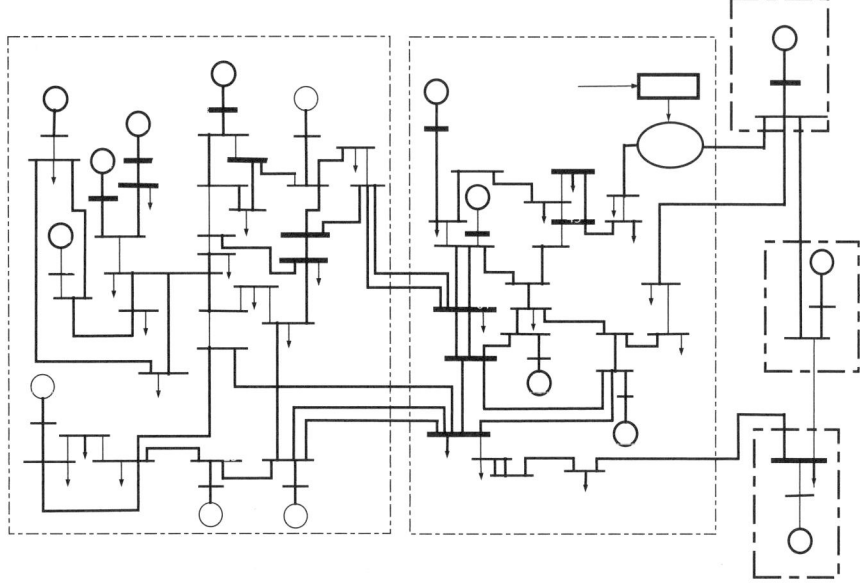

Figure 4.1. Sixteen machine five area study system with a FACTS device

4.1 Overview of the test system

A 16-machine, 68 bus test system is shown in Fig. 4.1. This is a reduced order equivalent of the interconnected New England test system (NETS) and New York power system (NYPS). There are five geographical regions out of which NETS and NYPS are represented by a group of generators whereas, import from each of the three other neighboring areas #3, #4 and #5 are approximated by equivalent generator models.

Generators G1 to G9 are the equivalent representation of the NETS generation whilst machines G10 to G13 represent the generation of the NYPS. Generators G14 to G16 are the dynamic equivalents of the three neighboring areas connected to the NYPS. There are three major transmission corridors between NETS and NYPS connecting buses #60-#61, #53-#54 and #27-#53. All these corridors have double-circuit tie-lines for which the line parameters are given Appendix A at the end of the book. In steady-state, the tie-line power exchange between NETS and NYPS is 700 MW in total.

The NYPS is required to import 1500 MW from Area #5. To facilitate this large amount of power transfer, either a series connected FACTS device, for example a thyristor controlled series capacitor (TCSC) or a thyristor controlled phase angle regulator (TCPAR) could be installed in the line connecting buses #18 and #50 or a shunt connected device such as an SVC could be installed at either of the buses. In this book, control design exercise has been carried out with different types of FACTS devices installed in the NYPS to Area #5 corridor.

4.2 Models of different components

As already discussed accurate modelling of the generators and their excitation systems is of fundamental importance for studying the dynamic behavior of power systems. Besides generators and excitation systems, other components such as the dynamic loads (e.g. induction motor type), controllable devices (e.g. thyristor controlled series capacitor (TCSC), power system stabilizer (PSS)), prime-movers etc. need to be modelled as well. The dynamic behavior of these devices are described through a set of differential equations but the power flow in the network is represented by a set of algebraic equations. This gives rise to a set of differential-algebraic equations (DAE) describing the power system behavior. Different types of models have been reported in the literature for each of the power system components depending upon their specific application [Kundur, 1994, Sauer and Pai, 1998]. In this section, the relevant equations governing the dynamic behavior of only the specific types of models used in this study is described. We have followed the IEEE recommended practice regarding d-q axis orientation [Concordia, 1969] of a synchronous generator. This results in a negative d axis component of stator current for overexcited generator delivering power to the system.

4.2.1 Generators

All the generators of the test system (G1 to G16) are represented by a sub-transient model [Sauer and Pai, 1998] with four equivalent coils on the rotor using the IEEE convention. Besides the field coil, there is one equivalent damper coil in the direct axis and two in the quadrature axis. The mechanical input power to the generator is assumed to be constant during the disturbances such as a 3-phase fault, obviating the need for modelling the prime-mover. The differential equations governing the sub-transient dynamic behavior of the generator i is given by:

$$\frac{d\delta_i}{dt} = \omega_i - \omega_s \tag{4.1}$$

$$\frac{d\omega_i}{dt} = \frac{\omega_s}{2H}[T_{mi} - D(\omega_i - \omega_s) - \frac{(X_{di}'' - X_{lsi})}{(X_{di}' - X_{lsi})}E_{qi}' I_{qi} - \frac{(X_{di}' - X_{di}'')}{(X_{di}' - X_{lsi})}\psi_{1di}I_{qi}$$
$$- \frac{(X_{qi}'' - X_{lsi})}{(X_{qi}' - X_{lsi})}E_{di}' I_{di} + \frac{(X_{qi}' - X_{qi}'')}{(X_{qi}' - X_{lsi})}\psi_{2qi}I_{di} + (X_{qi}'' - X_{di}'')I_{qi}I_{di}] \tag{4.2}$$

$$\frac{dE_{qi}'}{dt} = \frac{1}{T_{doi}'}[-E_{qi}' - (X_{di} - X_{di}')\{-I_{di} - \frac{(X_{di}' - X_{di}'')}{(X_{di}' - X_{lsi})^2}$$
$$(\psi_{1di} - (X_{di}' - X_{lsi})I_{di} - E_{qi}')\} + E_{fdi}] \tag{4.3}$$

$$\frac{dE_{di}'}{dt} = -\frac{1}{T_{qoi}'}[E_{di}' + (X_{qi} - X_{qi}')\{I_{qi} - \frac{(X_{qi}' - X_{qi}'')}{(X_{qi}' - X_{lsi})^2}$$
$$(-\psi_{2qi} + (X_{qi}' - X_{lsi})I_{qi} - E_{di}')\}] \tag{4.4}$$

$$\frac{d\psi_{1di}}{dt} = \frac{1}{T_{doi}''}[-\psi_{1di} + E_{qi}' + (X_{di}' - X_{lsi})I_{di}] \tag{4.5}$$

$$\frac{d\psi_{2qi}}{dt} = -\frac{1}{T_{qo}''}[\psi_{2qi} + E_{di}' - (X_{qi}' - X_{lsi})I_{qi}] \tag{4.6}$$

for $i = 1, 2, ..., m$, where,
m : total number of generators,
δ_i : generator rotor angle,
ω_i : rotor angular speed,
E_{qi}' : transient emf due to field flux-linkage,
E_{di}' : transient emf due to flux-linkage in q-axis damper coil,
ψ_{1di} : sub-transient emf due to flux-linkage in d-axis damper,
ψ_{2qi} : sub-transient emf due to flux-linkage in q-axis damper,
I_{di} : d-axis component of stator current,
I_{qi} : q-axis component of stator current,
$X_{di}, X_{di}', X_{di}''$: synchronous, transient and sub-transient reactances, respectively along d-axis,
$X_{qi}, X_{qi}', X_{qi}''$: synchronous, transient and sub-transient reactances, respectively along q-axis,
T_{do}', T_{do}'' : d-axis open-circuit transient and sub-transient time constants, respectively
T_{qo}', T_{qo}'' : q-axis open-circuit transient and sub-transient time constants, respectively

For stability studies, the stator transients are assumed to be much faster compared to the swing dynamics. Hence, the stator quantities are assumed to be related to the terminal bus quantities through algebraic equations rather than state equations. The stator algebraic equations are given by:

$$V_i \cos(\delta_i - \theta_i) - \frac{(X_{di}'' - X_{lsi})}{(X_{di}' - X_{lsi})}E_{qi}' - \frac{(X_{di}' - X_{di}'')}{(X_{di}' - X_{lsi})}\psi_{1di} + R_{si}I_{qi} - X_{di}''I_{di} = 0 \tag{4.7}$$

$$V_i \sin(\delta_i - \theta_i) + \frac{(X_{qi}'' - X_{lsi})}{(X_{qi}' - X_{lsi})}E_{di}' - \frac{(X_{qi}' - X_{qi}'')}{(X_{qi}' - X_{lsi})}\psi_{2qi} - R_{si}I_{di} - X_{qi}''I_{di} = 0 \tag{4.8}$$

for $i = 1, 2, ..., m$, where,
V_i : generator terminal voltage,
R_{si} : resistance of the armature,

X_{lsi} : armature leakage reactance.

The notation is standard as in [Sauer and Pai, 1998]. The parameters used for the test system are given in Appendix B at the end of the book.

4.2.2 Excitation systems

The generators G1 to G8 are equipped with slow excitation system (IEEE-DC1A) whilst G9 is equipped with a fast acting static excitation system (IEEE ST1A) and a speed-input power system stabilizer (PSS) [Kundur, 1994, Lee, 1992] to ensure adequate damping for its local modes. The rest of the generators are under manual excitation control.

The differential equations governing the behavior of an IEEE-DC1A type excitation system are given by:

$$\frac{dV_{tri}}{dt} = \frac{1}{T_{ri}}[-V_{tri} + V_{ti}] \quad (4.9)$$

$$\frac{dE_{fdi}}{dt} = -\frac{1}{T_{Ei}}[K_{Ei}E_{fdi} + E_{fdi}A_{ex}e^{B_{ex}E_{fdi}} - V_{ri}] \quad (4.10)$$

$$\frac{dV_{ri}}{dt} = \frac{1}{T_{Ai}}[\frac{K_{Ai}K_{Fi}}{T_{Fi}}R_{Fi} + K_{Ai}(V_{refi} - V_{tri}) - \frac{K_{Ai}K_{Fi}}{T_{Fi}}E_{fdi} - V_{ri}] \quad (4.11)$$

$$\frac{dR_{Fi}}{dt} = \frac{1}{T_{Fi}}[-R_{Fi} + E_{fdi}] \quad (4.12)$$

where,
E_{fdi} : field voltage,
V_{tri} : measured voltage state variable after sensor lag block,
and the rest of the notation carries their standard meaning [Sauer and Pai, 1998].

The governing equations for the IEEE-ST1A type excitation system are given by:

$$\frac{dV_{tri}}{dt} = \frac{1}{T_{ri}}[-V_{tri} + V_{ti}] \quad (4.13)$$

$$E_{fdi} = K_{ai}(V_{refi} - V_{tri}) \quad (4.14)$$

The fast acting static excitation system at generator G9 is equipped with power system stabilizers (PSSs) to provide supplementary damping control for the local modes. The feedback signal for this PSS is the measured speed of the generator G9. The dynamic response of the PSS is modelled by the following equation

$$V_{pssi} = K_{pssi} \frac{sT_w}{(1+sT_w)} \frac{(1+sT_{1i})}{(1+sT_{2i})} \frac{(1+sT_{3i})}{(1+sT_{4i})} \qquad (4.15)$$

where,
V_{ti} : measured terminal voltage,
and the rest of the notation carries their standard meaning as in [Sauer and Pai, 1998]. The parameters used for the test system are given in Appendix B at the end of the book.

4.2.3 Network power flow model

The network power balance equations pertaining to the generator buses are given by:

$$V_i \cos(\delta_i - \theta_i) I_{qi} - V_i \sin(\delta_i - \theta_i) I_{di} - S_{pi} = 0 \qquad (4.16)$$
$$-V_i \sin(\delta_i - \theta_i) I_{qi} - V_i \cos(\delta_i - \theta_i) I_{di} - S_{qi} = 0 \qquad (4.17)$$

where,

$$S_{pi} = \sum_{k=1}^{k=n} V_i V_k [G_{ik} \cos(\theta_i - \theta_k) + B_{ik} \sin(\theta_i - \theta_k)]$$

$$S_{qi} = \sum_{k=1}^{k=n} V_i V_k [G_{ik} \sin(\theta_i - \theta_k) - B_{ik} \cos(\theta_i - \theta_k)]$$

for $i = 1, 2, ..., m$

Power balance equations for the non-generator buses are given by:

$$P_{Li}(V_i) + \sum_{k=1}^{k=n} V_i V_k [G_{ik} \cos(\theta_i - \theta_k) + B_{ik} \sin(\theta_i - \theta_k)] = 0 \qquad (4.18)$$

$$Q_{Li}(V_i) + \sum_{k=1}^{k=n} V_i V_k [G_{ik} \sin(\theta_i - \theta_k) - B_{ik} \cos(\theta_i - \theta_k)] = 0 \qquad (4.19)$$

for $i = m+1, ..., n$
where, n is the total number of buses in the system and $Y_{ik} = G_{ik} + jB_{ik}$ is the element of the ith row and k^{th} column of the bus admittance matrix Y.

The load-flow data are given in the Appendix A at the end of the book.

4.3 Modelling of FACTS devices

Flexible AC transmission systems (FACTS) devices are installed in power systems to exert continuous control over the voltage profile or power flow pattern [Hingorani and Gyugyi, 2000]. They enable the voltage profile and power flows to be changed in such a way that thermal limits are not exceeded, stability margins are increased, losses minimized, contractual requirements fulfilled, etc. without violating the economic generation dispatch schedule [Noroozian et al., 1997b]. However, the sole presence of these devices does not improve the overall damping of the system appreciably. To enforce extra damping, supplementary control is required to be added to these FACTS devices and methods for designing such damping control through FACTS devices occupies a significant part of this book.

In this section, the steady-state and small-signal dynamic models of several series and shunt connected FACTS devices are briefly described. The power injection model is used for steady-state representation of these devices as it is most suitable for incorporation into an existing power flow algorithm without altering the bus admittance matrix Y [Noroozian et al., 1997a]. The power injection equations are given for different types of devices and the jacobian terms for these equations with respect to the state as well as the algebraic variables are presented in Appendix C at the end of the book. The small-signal dynamic models of the series connected devices are presented considering a single time constant block to represent the response time of the switching circuitry. For the shunt voltage control devices, a separate voltage control loop is involved with suitable response time of the voltage sensing hardware and time constants of the voltage regulator block.

4.3.1 Thyristor controlled series capacitor (TCSC)

A TCSC is a capacitive reactance compensator which consists of a series capacitor bank shunted by a thyristor controlled reactor (TCR) in order to provide a smooth variation in series capacitive reactance [Hingorani and Gyugyi, 2000, Song and Johns, 1999]. The circuit configuration of a typical TCSC is shown in Fig. 4.2.

When the TCR firing angle is 180 degrees, the reactor becomes non-conducting and the series capacitor has its normal impedance. As the firing angle is advanced from 180 degrees to less than 180 degrees, the capacitive impedance increases. On the other hand, when the firing angle is 90 degrees, the reactor become fully conducting and the TCSC helps in limiting the fault current [Hingorani and Gyugyi, 2000]. The control over firing angle produces a variable effective capacitance, which partly compensates for the transmission line in-

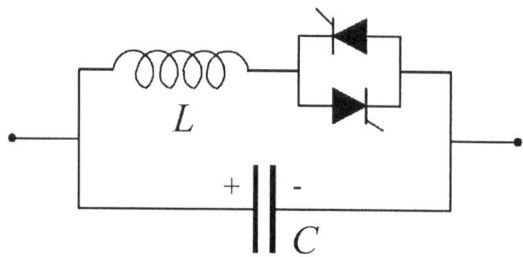

Figure 4.2. Thyristor controlled series capacitor (TCSC) topology

ductance and thereby, controls the power flow through the line.

The control action of the TCSC is expressed in terms of its percentage compensation k_c, defined as $k_c = \frac{X_C}{X_L} \times 100\%$ where X_L is the reactance of the line and X_C is the effective capacitive reactance offered by the TCSC.

Let us consider that the TCSC is connected in the line between bus k and m. In this case the resistance of the line is neglected for simplicity of calculation. If $\overline{I}$ is the current flowing through the line, the TCSC having capacitive reactance X_C can be represented by a voltage source $\overline{V}_{se}$ as shown in Fig. 4.3 where $\overline{V}_{se}$ is given by (4.20)

$$\overline{V}_{se} = -jX_C\overline{I} \qquad (4.20)$$

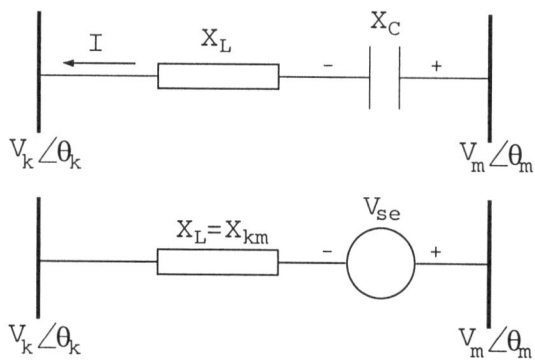

Figure 4.3. Voltage source model of TCSC

Test System Model

The injection model is obtained by replacing the voltage source by an equivalent current source $\overline{I}_s$ in parallel with the line as shown in Fig. 4.4 where $\overline{I}_s$ is given by (4.21)

$$\overline{I}_s = \frac{\overline{V}_{se}}{X_{km}} \quad (4.21)$$

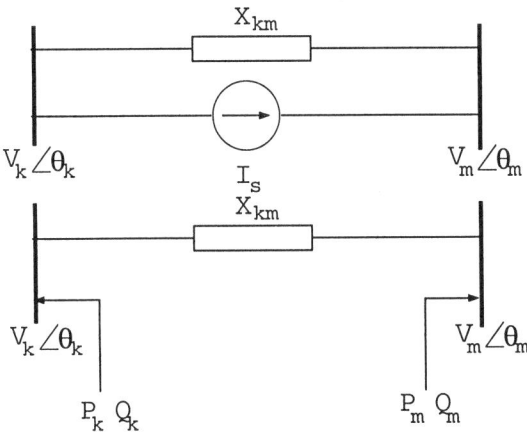

Figure 4.4. Power injection model of TCSC

The current source $\overline{I}_s$ corresponds to the injection powers $\overline{S}_k$ and $\overline{S}_m$ which are given by (4.22) and (4.23)

$$\overline{S}_k = \overline{V}_k \left(-\overline{I}_s\right)^* \quad (4.22)$$

$$\overline{S}_m = \overline{V}_m \left(\overline{I}_s\right)^* \quad (4.23)$$

From (4.22) and (4.23) the real and reactive power injection equations of the TCSC connected between bus k and m can be obtained by algebraic simplification and are given by (4.25)-(4.28) where percentage compensation (k_c) is given by (4.24).

$$k_c = \frac{X_C}{X_L} \quad (4.24)$$

and X_L is the reactance of the line.

$$P_k = \frac{k_c}{(k_c - 1)} V_k V_m B_{km} \sin\left(\theta_k - \theta_m\right) \quad (4.25)$$

$$Q_k = \frac{k_c}{(k_c - 1)} B_{km} \left[V_k^2 - V_k V_m \cos(\theta_k - \theta_m)\right] \quad (4.26)$$

$$P_m = \frac{k_c}{(k_c - 1)} V_m V_k B_{km} \sin(\theta_m - \theta_k) \quad (4.27)$$

$$Q_m = \frac{k_c}{(k_c - 1)} B_{km} \left[V_m^2 - V_m V_k \cos(\theta_m - \theta_k)\right] \quad (4.28)$$

The dynamic characteristics of the TCSC is assumed to be modelled by a single time constant ($T_{tcsc} = 0.02s$) representing the response time of the TCSC control circuit as follows:

$$\frac{d}{dt}\Delta k_c = \frac{1}{T_{tcsc}}\left(-\Delta k_c + \Delta k_{c-ref} + \Delta k_{c-ss}\right) \quad (4.29)$$

The small-signal dynamic model is given in Fig. 4.5 where, Δk_c is the incremental change in value of k_c about the nominal value of 0.5 (50% compensation). Δk_{c-ref} is the reference setting which is augmented by Δk_{c-ss} within a limit of $\Delta k_{c-max} = 0.3$ and $\Delta k_{c-min} = -0.4$ in the presence of supplementary damping control.

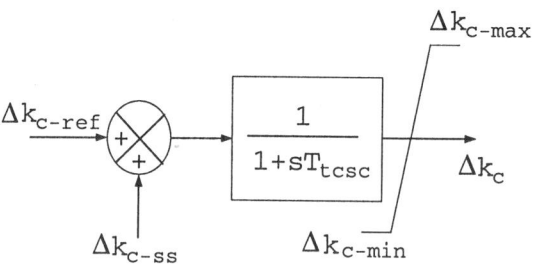

Figure 4.5. Small-signal dynamic model of TCSC

4.3.2 Static VAr compensator (SVC)

A static VAr compensator (SVC) is a shunt connected static VAr generator or absorber whose output is adjusted to exchange capacitive or inductive current so as to maintain or control specific variables of the electrical power system (typically bus voltage) [Hingorani and Gyugyi, 2000, Song and Johns, 1999]. A typical topology of a SVC comprises a parallel combination of a thyristor controlled reactor and a fixed capacitor as shown in Fig. 4.6.

The reactive power injection of a SVC connected to bus k is given by:

Test System Model

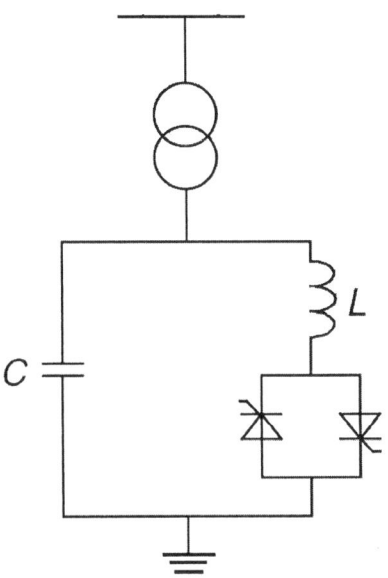

Figure 4.6. Static VAr compensator (SVC) topology

$$Q_k = V_k^2 B_{svc} \qquad (4.30)$$

where $B_{svc} = B_C - B_L$ and B_C and B_L are the susceptance of the fixed capacitor and thyristor controlled reactor respectively.

The small-signal dynamic model of a SVC is given in Fig. 4.7 where, T_{svc} is the response time of the switching circuitry, T_m is the time constant representing the delay in measurement and T_{v1} and T_{v2} are the time constants of the voltage regulator block. ΔB_{svc} is given by:

$$\Delta B_{svc} = \Delta B_C - \Delta B_L \qquad (4.31)$$

The dynamic equations are given by:

$$\frac{d}{dt}\Delta B_{svc} = \frac{1}{T_{svc}}\left[-\Delta B_{svc} + \left(1 - \frac{T_{v1}}{T_{v2}}\right)\Delta V_{r-svc} - \frac{K_v T_{v1}}{T_{v2}}\Delta V_{t-svc}\right]\ldots$$

$$+ \frac{K_v T_{v1}}{T_{v2} T_{svc}}[\Delta V_{ss-svc} + \Delta V_{ref}] \quad (4.32)$$

$$\frac{d}{dt}\Delta V_{r-svc} = \frac{1}{T_{v2}}\left(-\Delta V_{r-svc} - K_v \Delta V_{t-svc} + K_v V_{ref} + K_v V_{ss-svc}\right)$$
$$(4.33)$$

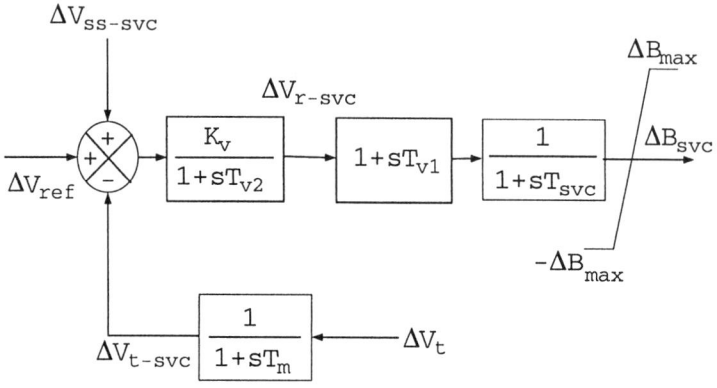

Figure 4.7. Small-signal dynamic model of SVC

$$\frac{d}{dt}\Delta V_{t-svc} = \frac{1}{T_m}(\Delta V_t - \Delta V_{t-svc}) \qquad (4.34)$$

Reference input ΔV_{ref}, in Fig. 4.7, is set to a point to maintain acceptable voltage at the SVC bus, while the supplementary input ΔV_{ss-svc} is controlled to damp inter-area oscillations. A thyristor controlled reactor (TCR) of 150 MVAr capacity is considered in parallel with a fixed capacitor (FC) of 200 MVAr. At 1.0 pu voltage, this corresponds to a susceptance range of -1.50 pu to 2.0 pu which sets the limits of the SVC output. The steady-state settings of the FC and the TCR are 150 MVAr and 33 MVAr, respectively.

4.3.3 Thyristor controlled phase angle regulator (TCPAR)

A thyristor controlled phase angle regulator (TCPAR) is basically a phase-shifting transformer adjusted by thyristor switches to provide a rapidly variable phase angle [Hingorani and Gyugyi, 2000, Song and Johns, 1999]. In general, phase shifting is obtained by adding a perpendicular voltage vector in series with a phase. This vector, which can be made variable using a number of power electronics topologies, is derived from the other two phases via a shunt connected transformer [Hingorani and Gyugyi, 2000, Song and Johns, 1999] as shown in Fig. 4.8.

A TCPAR connected in the line between bus k and m as shown in Fig. 4.9 with its exciter transformer being fed from the bus k side. The injected voltage can be modelled as an ideal voltage source $\overline{V}_{se} = V_s \angle \phi$ in series with the line impedance $\overline{Z}_{km}$. The injection model is obtained by replacing the voltage

Test System Model

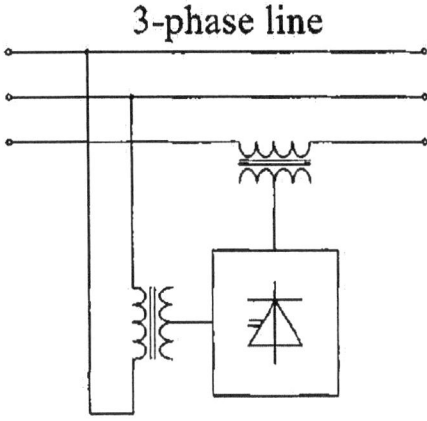

Figure 4.8. Thyristor controlled phase angle regulator (TCPAR) topology

source by an equivalent current source $\bar{I}_{se}$ in parallel with the line as shown in Fig. 4.9 where $\bar{I}_{se}$ and $\bar{I}_{sh}$ are given by:

$$\bar{I}_{se} = \frac{\bar{V}_{se}}{\bar{Z}_{km}} \tag{4.35}$$

$$\bar{I}_{sh} = \bar{I}_k - \bar{I}_{(se)} \tag{4.36}$$

The series current source $\bar{I}_{se}$ in addition to the shunt current $\bar{I}_{sh}$ corresponds to the injection powers $\bar{S}_k$ and $\bar{S}_m$ which are given by:

$$\bar{S}_k = \bar{V}_k\left(-\bar{I}_{sh} - \bar{I}_{se}\right)^* \tag{4.37}$$

$$\bar{S}_m = \bar{V}_m\left(\bar{I}_{se}\right)^* \tag{4.38}$$

From (4.37) and (4.38) the real and reactive power injection equations of the TCPAR connected between bus k and m can be obtained by algebraic simplification and are given by:

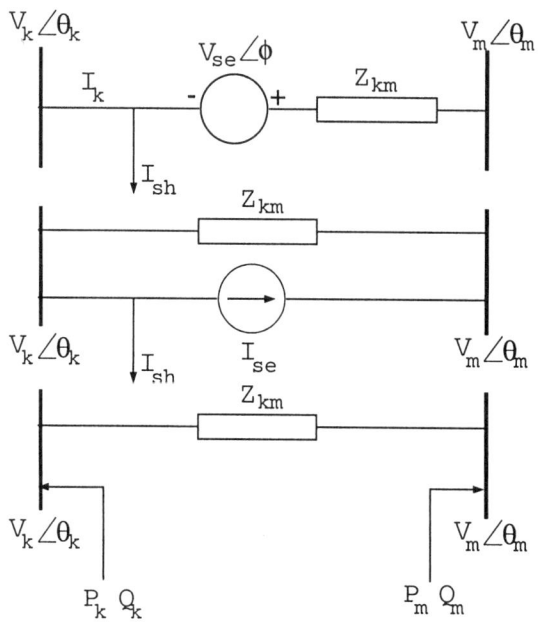

Figure 4.9. Power injection model of TCPAR

$$P_k = V_k V_m \left[G_{km} \{\cos\theta_{km} - \cos(\theta_{km} + \phi)\} + B_{km} \{\sin\theta_{km} - \sin(\theta_{km} + \phi)\} \right] \tag{4.39}$$

$$Q_k = V_k V_m \left[G_{km} \{\sin\theta_{km} - \sin(\theta_{km} + \phi)\} - B_{km} \{\cos\theta_{km} - \cos(\theta_{km} + \phi)\} \right] \tag{4.40}$$

$$P_m = V_m V_k \left[G_{mk} \{\cos\theta_{mk} - \cos(\theta_{mk} - \phi)\} + B_{mk} \{\sin\theta_{mk} - \sin(\theta_{mk} - \phi)\} \right] \tag{4.41}$$

$$Q_m = V_m V_k \left[G_{mk} \{\sin\theta_{mk} - \sin(\theta_{mk} - \phi)\} - B_{mk} \{\cos\theta_{mk} - \cos(\theta_{mk} - \phi)\} \right] \tag{4.42}$$

The small-signal dynamic model of a controllable phase shifter is given in Fig. 4.10 where T_{tcpar} represent the response time of the switching circuitry. The dynamic equation is given by:

$$\frac{d}{dt}\Delta\phi = \frac{1}{T_{tcpar}}\left(-\Delta\phi + \Delta\phi_{ref} + \Delta\phi_{ss}\right) \tag{4.43}$$

Reference input $\Delta\phi_{ref}$, in Fig. 4.10, is set to a point to ensure the desired power flow in the line, while the supplementary input $\Delta\phi_{ss}$ is controlled to damp inter-area oscillations. The steady state value of the phase angle ϕ was

Test System Model

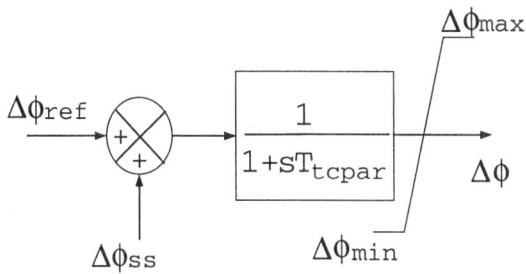

Figure 4.10. Small-signal dynamic model of TCPAR

set to 10 degrees for the required power transfer between area #5 and NYPS in the test system model.

4.4 Linearized system model

In general, the dynamic behavior of the power system is governed by the set of differential-algebraic equations (DAE) described in the earlier sections. This set of differential-algebraic equations is linearized with respect to the equilibrium point (x_0, y_0) as shown in Chapter 3.

The set of DAE equations needs to be reduced to a set of ordinary differential equations (ODE) which is a much more convenient form for analysis and design. To obtain the state-space representation of the test system, it is also required to have the input matrix B and the output matrix C as well as the system matrix A. The input matrix B has the same number of rows as that of A with only one non-zero entry corresponding to the states of the respective FACTS device, the rest of the elements being zero. The C matrix depends on how the feedback signals are related to the state and/or algebraic variables. The selection of the feedback signals is discussed in the next section. The entries of the D matrix are zero as there is no direct influence of the control input on the FACTS device by the measured signals. The A, B, C and D matrices represent the linearized model of the test system which is used for analysis and control design throughout the rest of this book. For details of the methodology for obtaining a linearized power system model, the readers are recommended to go through [Sauer and Pai, 1998].

If the specific case with one TCSC installed in the system is considered, the total number of state variables for the linearized system model is 132. By computing the eigen-values of the linearized system model, it is found that the system has four inter-area modes which are lightly damped as shown in Table 4.1. Out of the four inter-area modes, the first three are critical requiring

Table 4.1. Inter-area modes of the test system with TCSC

Eigen-value $\sigma \pm j\omega$	Damping ratio $\frac{-\zeta}{\sqrt{\zeta^2+\omega^2}}$	Frequency (Hz) $\frac{\omega}{2\pi}$
$-0.154 \pm j\, 2.46$	0.0626	0.3913
$-0.139 \pm j\, 3.19$	0.0435	0.5080
$-0.217 \pm j\, 3.92$	0.0554	0.6232
$-0.248 \pm j\, 4.97$	0.0499	0.7915

Table 4.2. Inter-area modes of the test system with SVC

Eigen-value $\sigma \pm j\omega$	Damping ratio $\frac{-\zeta}{\sqrt{\zeta^2+\omega^2}}$	Frequency (Hz) $\frac{\omega}{2\pi}$
$-0.154 \pm j\, 2.42$	0.0635	0.3853
$-0.137 \pm j\, 3.17$	0.0432	0.5039
$-0.218 \pm j\, 3.90$	0.0558	0.6204
$-0.248 \pm j\, 4.97$	0.0499	0.7913

additional damping. Mode #4, on its own, settles in less than 10 seconds as its frequency (0.79 Hz) is comparatively higher than the other modes (the higher the oscillation frequency, the faster is the settling for a given damping ratio). Since the influence of this mode on inter-area oscillation does not last beyond 10 seconds and an overall system settling time of 10-12 seconds is acceptable, it is not required to provide additional damping to this mode. Therefore, the objective is to design a controller to produce robust damping for the three critical inter-area modes.

With an SVC installed in the system, the total number of states is 136. The inter-area modes of the system with a SVC installed is shown in Table 4.2.

The total number of states of the system with the TCPAR installed is 134. The inter-area modes of the system with a TCPAR installed is shown in Table 4.3.

4.5 Choice of remote signals

The feedback stabilizing signals are chosen based on the modal controllability observability and residue analysis as described earlier in Chapter 3.

Test System Model   55

Table 4.3. Inter-area modes of the test system with TCPAR

Eigen-value $\sigma \pm \jmath\omega$	Damping ratio $\frac{-\zeta}{\sqrt{\zeta^2+\omega^2}}$	Frequency (Hz) $\frac{\omega}{2\pi}$
$-0.155 \pm \jmath\, 2.40$	0.0644	0.3813
$-0.138 \pm \jmath\, 3.13$	0.0439	0.4988
$-0.219 \pm \jmath\, 3.89$	0.0561	0.6187
$-0.248 \pm \jmath\, 4.96$	0.0499	0.7896

Table 4.4. Normalized residues for active power flow signals from different lines with TCSC installed in the system

Mode 1		Mode 2		Mode 3	
Line	Residue	Line	Residue	Line	Residue
51 - 45	1.00	**18 - 16**	1.00	**13 - 17**	1.00
50 - 51	0.67	41 - 14	0.79	36 - 17	0.67
50 - 18	0.63	42 - 18	0.75	60 - 61	0.34
35 - 34	0.63	41 - 42	0.55	53 - 30	0.30
45 - 35	0.63	53 - 30	0.34	61 - 36	0.29
34 - 36	0.57	53 - 47	0.31	54 - 53	0.29
53 - 47	0.56	51 - 45	0.29	**50 - 51**	0.27
36 - 34	0.55	**50 - 51**	0.29	**50 - 18**	0.25
53 - 54	0.50	36 - 17	0.28	34 - 36	0.24
41 - 40	0.49	**50 - 18**	0.28	68 - 37	0.21

For the test system with a TCSC installed, the active power flow in the transmission lines was chosen as the feedback signal. Bus voltage and reactive power flow signals were not chosen, as besides oscillatory modes, the exciter and flux-decay dynamics are also dominant in those signals. The machine speed signal requires an additional phase lead requirement of 90 degrees as compared to the active power signal and the latter is, therefore, considered. The normalized residues for different signals corresponding to each critical inter-area mode of the test system is shown in Table 4.4. Only a few of the most effective signals with maximum residues for each mode are shown in the following tables.

The results reveal that $P_{51,45}$, $P_{18,16}$ and $P_{13,17}$ are the most effective signals for mode #1, mode #2 and mode #3, respectively, where $P_{51,45}$, $P_{18,16}$ and $P_{13,17}$ indicate the power-flow in the lines connecting buses #51-#45, buses #18-#16

Table 4.5. Normalized residues for active power flow signals from different lines with SVC installed in the system

Mode 1		Mode 2		Mode 3	
Line	Residue	Line	Residue	Line	Residue
13 - 17	1.00	**18 - 16**	1.00	13 - 17	1.00
51 - 45	0.74	13 - 17	0.81	17 - 36	0.64
50 - 51	0.62	18 - 42	0.78	60 - 61	0.35
18 - 69	0.60	41 - 42	0.56	53 - 30	0.30
35 - 34	0.60	53 - 30	0.33	61 - 36	0.30
45 - 35	0.60	47 - 53	0.32	54 - 53	0.29

and buses #13-#17, respectively. It is evident from the table that that although mode #1 is noticeably observable in the locally available signals $P_{50,51}$, $P_{50,18}$ (shown in boldface), the observability of the other two modes are poor. Therefore, it is concluded that local signals are not always the most appropriate for all the critical modes. Under these circumstances, signals from remote locations need to be transmitted to the controller site.

The normalized residues for power signals from different lines with the SVC installed in the system is shown in Table 4.5. It can be seen that in this case, only two signals $P_{13,17}$ and $P_{18,16}$ are required for observing all the three inter-area modes.

With the TCPAR installed in the system, $P_{51,45}$, $P_{14,41}$ and $P_{13,17}$ are the most appropriate feedback signals for the three dominant inter-area modes as shown in Table 4.6. It can be noted that for mode#2, $P_{18,42}$ is almost as good as $P_{14,41}$.

Considering the appropriate feedback signals as determined above, the output matrix C for the state-space model of the test system is constructed with each FACTS device installed in the system [Sauer and Pai, 1998].

4.6 Simplification of system model

The order of the controllers synthesized using $\mathcal{H}_\infty$ based techniques are at least as high as the order of the open-loop system (134 in case of the present test system with TCSC). It might be even higher with the incorporation of the weighting functions. Therefore, it is mandatory to simplify the system model, if possible, to ease the design procedure and avoid complexity in the final con-

Table 4.6. Normalized residues for active power flow signals from different lines with TCPAR installed in the system

Mode 1		Mode 2		Mode 3	
Line	Residue	Line	Residue	Line	Residue
51 - 45	1.00	**14 - 41**	1.00	**13 - 17**	1.00
50 - 51	0.83	18 - 42	0.98	36 - 17	0.66
34 - 35	0.82	41 - 42	0.69	60 - 61	0.35
35 - 45	0.82	53 - 30	0.40	53 - 30	0.30
18 - 50	0.79	47 - 53	0.40	61 - 36	0.30
34 - 36	0.75	41 - 40	0.34	54 - 53	0.29
47 - 53	0.70	40 - 48	0.33	34 - 36	0.23

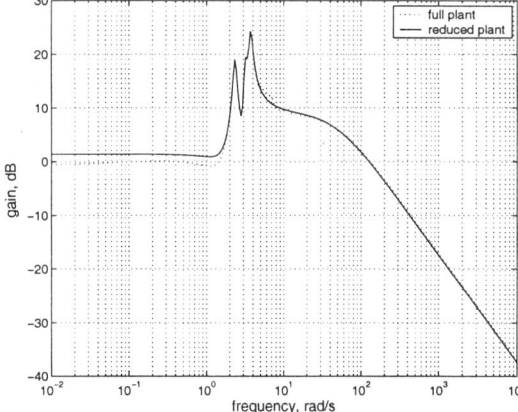

Figure 4.11. Frequency response of original and simplified system with TCSC troller.

The 132 order test system model could be simplified down to a 7th order reduced equivalent without loosing much information in the relevant frequency range of interest (0.1 - 1.0 Hz). This is done by starting with a higher order simplified model and going down to point where the frequency response of the original and simplified model starts showing some differences. The Hankel singular values (HSVs), discussed in the earlier chapter, give an idea about the extent of simplification that can be achieved. The frequency response of the original and the simplified system, shown in Fig. 4.11, confirms that they are closely matched in the frequency range of interest.

The model reduction exercise has been carried out using the 'schmr' function available in the robust control toolbox [Packard et al., 2002] of Matlab [mat, 1998]. For other types of FACTS devices installed in the system, the reduced system model for control design is derived similarly.

Throughout this book, the control design methodologies are illustrated using the study system model with TCSC installed. The design methods are general and have been applied for control design with other types of FACTS devices installed in the system. Simulation results are presented for different types of FACTS devices to highlight the generic nature of the control design procedures.

References

[mat, 1998] (1998). *Matlab Users Guide*. The Math Works Inc., USA.

[Concordia, 1969] Concordia, C. (1969). Eee committee report on recommended phasor diagram for synchronous machines. *IEEE Transactions on Power Apparatus and Systems*, 88(11):1593–1610.

[Hingorani and Gyugyi, 2000] Hingorani, N.G. and Gyugyi, L. (2000). *Understanding FACTS*. IEEE Press, USA.

[Kundur, 1994] Kundur, P. (1994). *Power System Stability and Control*. McGraw Hill, USA.

[Lee, 1992] Lee, D.C. (1992). *IEEE recommended practice for excitation system models for power system stability studies*. Energy development and power generation committee of power engineering society.

[Noroozian et al., 1997a] Noroozian, M., Angquist, L., Gandhari, M., and Andersson, G. (1997a). Improving power system dynamics by series-connected FACTS devices. *IEEE Transactions on Power Delivery*, 12(4):1635–1641.

[Noroozian et al., 1997b] Noroozian, M., Angquist, L., Gandhari, M., and Andersson, G. (1997b). Use of UPFC for optimal power flow control. *IEEE Transactions on Power Delivery*, 12(4):1629–1634.

[Packard et al., 2002] Packard, A., Balas, G.J., Safonov, M., and R., Chiang (2002). *Robust Control Toolbox for use with Matlab*. The Math Works Inc., USA.

[Sauer and Pai, 1998] Sauer, P.W. and Pai, M.A. (1998). *Power System Dynamics and Stability*. Prentice Hall, USA.

[Song and Johns, 1999] Song, Y.H. and Johns, A.T. (1999). *Flexible AC Transmission Systems*. IEE Power and Energy series, UK.

Chapter 5

POWER SYSTEM STABILIZERS

5.1 Introduction

Back in the 1940s and 50s, considerable emphasis was placed on the economic design of generators, especially, those of large ratings. This resulted in machines with large steady-state synchronous reactances,and resulting in poor load-voltage characteristics, especially when connected through long transmission lines. On load, the armature reaction effect caused a reduction of field flux resulting in significant drop in the overall synchronizing torque. Therefore, the transient stability related problems for synchronous operation became the major concern. The problem was resolved by using high gain,fast acting excitation control systems. Bus-fed thyristor-controlled static excitation was increasingly used such that the voltage regulator provided sufficient synchronizing torque by virtually eliminating the effect of armature reaction on reduction in synchronizing torque. However, voltage regulator action was found to introduce negative damping torque at high power output and weak external network conditions represented by long overhead transmission lines, a very common operating situation in power systems in US and Canada. Negative damping gave rise to an oscillatory instability problem. The conflicting performance of the excitation control loop was resolved by modulating the voltage regulator reference input through an additional signal,which was meant to produce positive damping torque. The control circuitry producing this signal was termed a power system stabilizer (PSS).

With this brief genesis of PSS, we will focus on the factors influencing PSS design and practical implementation in a single machine infinite bus (SMIB) set up. The conceptual understanding of PSS in SMIB will be extended to the multi-machine case later. This chapter contains an account of various methods

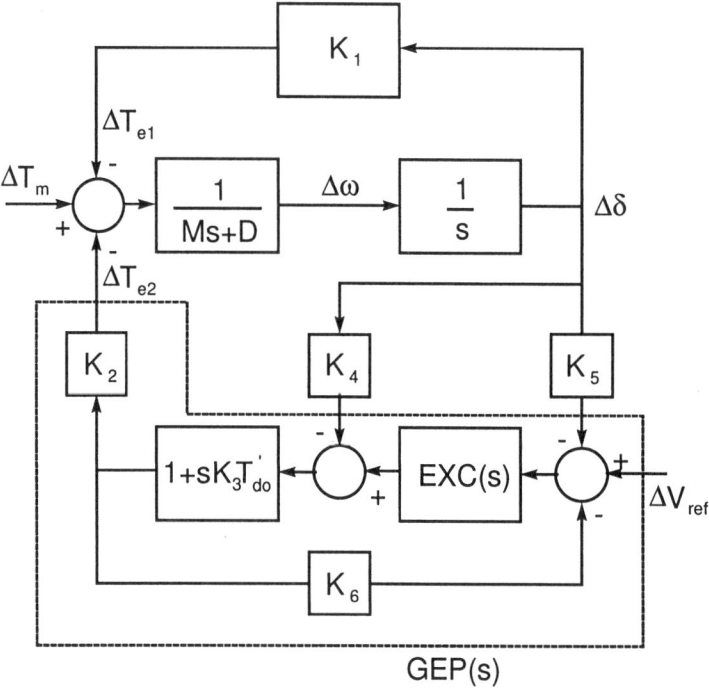

Figure 5.1. Heffron-Phillips block diagram of single machine infinite bus model

suggested or used to tune and coordinate PSS parameters for robust performance.

5.2 Basic Concept of PSS

The primary objective of a PSS is to introduce a component of electrical torque in the synchronous machine rotor that is proportional to the deviation of the actual speed from the synchronous speed. When the rotor oscillates, this torque acts as a damping torque to counter the oscillation.

To understand the operation of a PSS, consider the block diagram in Fig. 5.1. This diagram was developed by Heffron and Phillips [Heffron and Phillips, 1952] to represent the dynamics of a single synchronous generator connected to the grid through a line. The simple electromechanical generator model with mechanical swing and field flux dynamics was used. The dynamics of the voltage regulator was represented as a transfer function block $EXC(s)$. De Mello and Concordia adopted this model in their paper [de Mello and Concordia, 1969] to develop an understanding of the mechanism of oscillations in terms of

Power System Stabilizers

damping and synchronizing torque.

The parameters $K_1 - K_6$ in Fig. 5.1 are constants for a particular operating point but vary with the power output and the strength of the electrical network connecting the machine to the infinite bus. It is easy to identify several loops through which overall electrical torque is generated. In the absence of voltage regulator effect, the total electrical torque $\Delta T_e = \Delta T_{e1} + \Delta T_{e2}$ is

$$\Delta T_e = K_1 - \frac{K_2 K_3 K_4}{1 + sK_3 T'_{do}} \tag{5.1}$$

Thus the effect of the armature reaction is to reduce the synchronizing torque by $K_2 K_3 K_4$. The effect of armature reaction is to introduce positive damping, small as it may be. When the voltage regulator, with gain K_A and time constant T_A, is connected the drop in synchronizing torque due to demagnetizing action is scaled down by a factor of $K_A K_3 K_6$. The machine angle variation $\Delta \delta$ is amplified by the voltage regulator to produce a small negative synchronizing torque at frequencies much below $\frac{1}{T'_{do} T_A}$. This condition usually occurs when the machine is moderately loaded by connecting to a strong network where, K_1 is high and hence, the overall synchronizing torque is positive. The expression for damping torque as a function of the frequency ω is given by:

$$\Delta T_D = \frac{K_2 K_A K_5 \left(T_A/K_3 + T'_{do}\right) \omega}{\left(1/K_3 + K_6 K_A - \omega^2 T'_{do}\right)^2 + \left(T_A/K_3 + T'_{do}\right)^2 \omega^2} \Delta \delta \tag{5.2}$$

ΔT_D is negative for negative K_5 as all the other coefficients in (5.2) are positive. K_5 is negative for high loading and/or a weak network condition

The action of the PSS is effective through the transfer function block $GEP(s)$ between the electric torque and the reference voltage input with variation in the machine speed assumed to be zero. The expression for the transfer function $GEP(s)$ can be derived from the block diagram :

$$GEP(s) = \frac{K_2 K_3 EXC(s)}{(1 + sT'_{do} K_3) + K_3 K_6 EXC(s)} \tag{5.3}$$

The transfer function of the excitation system, $EXC(s)$, can be of any type. Here the excitation system is assumed to be of the static type with high gain and extremely low response time. The transfer function $GEP(s)$ can be obtained from a field test on the generator. It was shown in [Larsen and Swann, 1981, Padiyar, 1996] that for a very small value of K_5, $GEP(s)$ could be related to the closed-loop frequency response of the voltage regulator loop (with $\Delta \omega = 0$) as:

$$\frac{\Delta V_t(s)}{\Delta V_s(s)} = GEP(s) \left[\frac{K_6}{K_2} - \frac{K_5}{Ms^2 + K_e(s)} \right] \qquad (5.4)$$

$K_e(s)$ is the effective synchronizing torque. The transfer functions $GEP(s)$ and $\frac{\Delta V_t(s)}{\Delta V_s(s)}$ are very similar except around the local mode frequency, where they differ by about 180 degrees.

As system operating conditions change, the gain and phase characteristics of the transfer function $GEP(s)$ change. Ideally, the PSS transfer function should be reciprocal of that for providing a prescribed amount of damping with speed input. However, that would result in an improper transfer function for the required compensator, not feasible in practice as it amplifies the noise mixed with the signal. A practical approach is to have a phase lead-lag circuit that provides adequate compensation over the desired frequency range. A gain is selected to provide adequate damping, primarily for the local modes but also for the inter-area modes if the unit participates in them.

5.3 Stabilizing signals for PSS

The choice of the stabilizing signal for a PSS is influenced by many factors. The signals should be locally available, be easily measured or synthesized and have a very high signal to noise ratio [Larsen and Swann, 1981]. The most widely used signal is the rotor speed. However bus frequency, electrical power, accelerating power and synthesized rotor speed are also used.

Rotor speed is sensitive to noise and torsional interaction in thermal units. The solution is to locate the speed sensor on the shaft where the node of the first torsional mode occurs. However, this location may not be accessible as it might occur in the middle of the shaft casing. Usually, a filter is used to reduce the interaction but this filter is prone to introduce a destabilizing effect on the exciter mode that increases with increasing stabilizer gain. The generator rotor speed is also sensitive to intra-plant modes (1.5 to 2.5 Hz) and hence is not preferred as a PSS signal for a multi-unit power plant complex.

The frequency signal is insensitive to an intra-plant mode. It can enhance the damping of the inter-area mode better than the speed signal as frequency is more sensitive to modes of oscillation between large areas than to the modes involving individual machines. The sensitivity of the frequency increases under weak external network conditions making it particularly useful for providing damping in these conditions. However, the frequency signal can be heavily corrupted with noise on the power system such as that of a nearby electric-arc furnace. During a rapid transient, the frequency signal undergoes a sudden phase shift causing a spike in the field voltage that gets reflected into the generator output voltage. Thus the frequency signal also needs a torsional filter although its duty is not as heavy as for a speed signal.

An accelerating power signal has the advantage of being immune to very low level torsional interactions [Bayne et al., 1977, de Mello et al., 1978]. However, the difficulty in obtaining the accelerating power signal is to account for variation in the mechanical power input. Ignoring changes in the mechanical input is not justified when loading on the unit changes. The output of the PSS, in such a situation, causes an undesirable excursion of output voltage and reactive power. This problem can be addressed using a Delta-P-Omega stabilizer [Lee et al., 1981]. An equivalent speed (ω_{eq}) is obtained from the integral of the difference between the deviation in mechanical (ΔP_m) and electrical power (ΔP_e). The deviation of the electrical power is measured directly but the deviation in the mechanical power is computed from the integral of the measured electrical output power and the measured speed passed through a filtered speed signal. Usually, the torsional modes are greatly attenuated in the electrical power signal and do not require an additional torsional filter. This prevents the exciter mode from becoming destabilized. A simple and standard torsional filter is sufficient to obtain the speed signal which allows manufacturers to go for a standard design of PSS, irrespective of the general torsional characteristics [Kundur, 1994].

5.4 Structure of PSS

The most commonly used structure of a PSS is shown in Fig. 5.2. This comprises a gain, phase compensation blocks, a washout filter, torsional filters when there are speed and frequency inputs and output limiters.

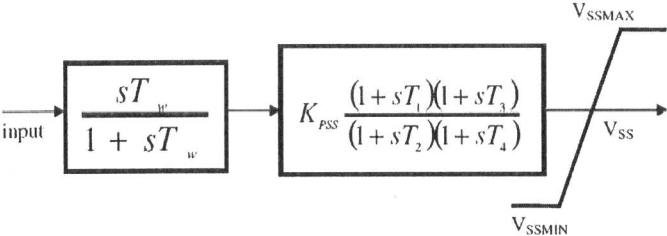

Figure 5.2. A commonly used structure of PSS

Washout circuit The washout circuit is a high-pass filter that prevents any steady change in speed, frequency and power affecting the field voltage because of the PSS action. The PSS is expected to act only during transient variation in the stabilizing signal. The value of T_w is chosen to give a band-pass effect to the signals containing local and inter-area modes. For local modes, T_w=1-2 s is satisfactory. A longer T_w, in the range of 10-20 s, is necessary for damping the inter-area modes. A noticeable improvement in the first swing stability is achieved with T_w set to 10.0 s. Higher T_w results in better terminal voltage response [Kundur et al., 1989].

Phase compensation block The primary objective of the phase compensation circuit is to counter the phase lag introduced in the transfer function,$GEP(s)$. In order to achieve only a pure damping torque contribution from the PSS, the phase compensator must cancel that phase-lag. It has already been mentioned that a pure phase-lead block is difficult to achieve in practice and hence a phase-lead-lag block is usually designed. The time constants T_1, T_2, T_3 and T_4 are chosen based based on the following criteria: [Larsen and Swann, 1981, Padiyar, 1996]

- The compensated phase i.e. the phase of the product of the two transfer functions $GEP(s)$ and $GPSS(s)$ should pass through $90°$ beyond 3.5 Hz for speed and power input and 2.0 Hz for frequency input. This is to avoid destabilizing the intra-plant mode.

- The compensated phase around the local mode should be between $20°-40°$ lag. This can be achieved by reducing the washout time-constant T_w. However, too small a T_w will not be effective in the inter-area mode frequency range, hence a compromise is needed.

- The high frequency gain of the compensator T_1T_3/T_2T_4 should be limited so as not to amplify the noise. The number of lead-lag blocks varies between one to three depending on the amount of phase compensation required.

Ideally a compensator should be obtained that maximizes bandwidth within which phase lag remains less than $90°$, because beyond $90°$ the damping will decrease with increasing gain. The frequency at which the compensated phase reaches $90°$, is dictated by the compensator center frequency $f_c = 1/2\pi\sqrt{T_1T_2}$. Greater f_c results in greater bandwidth but better damping for the local mode is achieved as long as $f_c < 5.0$ Hz. Usually, the frequency f_c that produces maximum local mode damping also contributes positive damping to the inter-area mode. However, the maximum damping to an inter-area mode will result for a f_c around 5.0 Hz [Larsen and Swann, 1981].

The gain and phase-lag of the transfer function $GEP(s)$ varies significantly with changing operating conditions. The gain increases as the generator power output increases and approaches a maximum under the strongest network conditions. The phase of the regulator loop gain is also maximum at this condition. Higher loop gain is better from the perspective of performance but is not so from the stability margin requirement. This establishes the maximum permissible gain of the stabilizer with maximum phase compensation requirement. For a stabilizer utilizing speed or power as input signals, the PSS must be tuned under these conditions. At a weak network conditions and reduced unit output, where the loop gain is low, the stability margin is not a problem. However, the the response time is longer.

The sensitivity of the frequency is high under weak network conditions, and the PSS should be tuned to produce maximum performance with a frequency input PSS. The most effective gain is selected with the help of the root-locus method to produce acceptable damping. Once a proper phase compensation is provided, the gain is varied to see how the local mode moves in the complex plane. Initially, the local mode moves towards the left and the damping increases. However, at high gain, the exciter mode moves to the right and crosses the imaginary axis. It is shown in [Larsen and Swann, 1981] that with a speed input PSS and a wide range of lead-lag setting, the damping of the most weakly damped mode has a maximum at about $1/3^{rd}$ of the gain at which stability is lost. In a power input PSS, this occurs at about $1/8^{th}$ of the instability gain. Optimal damping is achieved at about $2/3^{rd}$ of the gain [Larsen and Swann, 1981]. These values may not be optimal as the error in approximating $GEP(s)$ with the measured voltage regulator transfer function is significant, especially in the local mode frequency range. This conclusion is not valid for all types of excitation system and the gain must be such that it does not interfere with the damping of other modes and does not lead to excessive amplification of signal noise.

The output of a PSS should be limited otherwise, it might prevent the action of the AVR, particularly This situation might arise during load rejection. The AVR acts to reduce the terminal voltage, where rising rotor speed or bus frequency would drive the PSS to produce positive output. During a fault, however, a positive contribution from the PSS improves the transient stability in the first swing. A limit of 0.1 to 0.2 p.u. is normal. Excessive terminal voltage excursion beyond the range of 1.10 to 1.15 p.u. is controlled by another circuit called the terminal voltage limiter [Kundur, 1994]. On the negative side, the limit is set to -0.05 to -0.10 p.u. This limit is very important during the back swing after initial acceleration when the unit requires large synchronizing torque at high load angle to give a return to equilibrium in the post-fault state. A high limit may lead to a unit tripping if the PSS output is held to its negative limit due to a failure in the PSS.

The general damping and synchronizing torque concept is normally applied to a simple machine model with field circuit dynamics only. If the effect of the rotor damper winding is included, the shape of the transfer function $GEP(s)$ and the phase compensation requirement is different. Fig. 5.3 shows machine speed response in a typical SMIB system with and without a PSS when the machine is subjected to a 3-phase fault at its terminal for 80 ms. The PSS used in that simulation was the IEEE Type PSS2A the detail of which can be found in [Lee, 1992]. It can be inferred from the response that the PSS produced additional damping to settle the oscillations quickly.

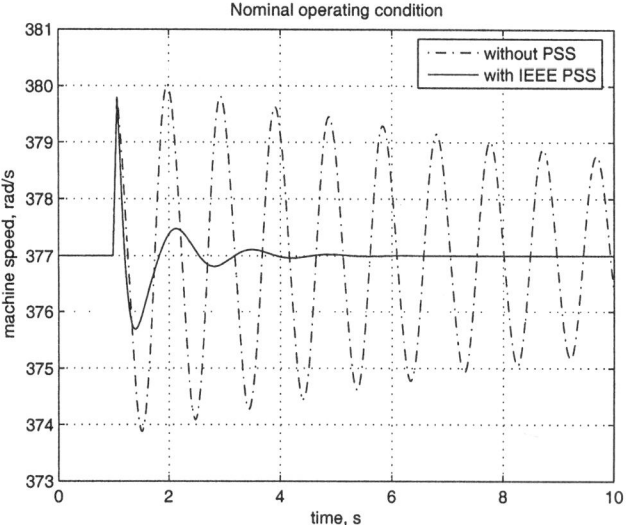

Figure 5.3. Response of a typical SMIB system under disturbance

5.5 Methods of PSS design

The design of a PSS through the damping and synchronizing torque concepts considers one machine at a time and is known as the SMIB approach. In reality, a large number of machines interact with each other through network voltages and power flows. The effect of the dynamics of other machines, therefore, must be taken into consideration whilst designing a particular PSS. This is known as coordinated tuning of PSS in a multi-machine system. A large number of papers have been written describing various methodologies for PSS design in a multi-machine system [Martins, 2000]. Broadly they fall into damping torque, frequency response and eigen value based and state space based categories. Frequency response techniques and the eigenvalue approach from state space representation of a system are equivalent from a control engineer's perspective.

However, for a SISO system, the frequency response approach reveals more interesting interpretations. Damping torque analysis is another level of simplification of the frequency domain approach where the analysis relates to the physics of the system. There have been many techniques that use a combination of these approaches.

5.5.1 Damping torque approach

The Heffron-Phillips constants $K_1 - K_6$ were developed for a multi-machine system in [Gooi et al., 1981]. The effect of other machines on a particular machine was represented as an additional equivalent synchronizing torque and damping torque co-efficient with the help of Mason's gain formula [Kuo and Golnaraghi, 2003]. A somewhat similar approach was reported in [Lim and Elangovan, 1985] but without Mason's gain formula. The damping torque concept, introduced in [de Mello and Concordia, 1969] and further applied in [Larsen and Swann, 1981], was extended to a multi-machine system in [Gibbard, 1988]. At selected operating conditions, the small signal analysis was used to examine the damping behavior of each of the machines, in the interarea mode frequency range. The transfer function $H_{vi}(s) = \frac{\Delta T_{ei}(s)}{\Delta V_{ssi}(s)}$ between the voltage reference input and the electrical torque output for each of the generators was computed. The effect of variation of the machine speed on electrical torque was modelled through two transfer functions: one through the rotor angle path as influenced by the terminal voltage variation effected through the network $H_{\delta i}(s) = \frac{\Delta T_{ei}(s)}{\Delta \omega_i(s)}$; the other through the speed input PSS, if present $H_{\omega i}(s) = \frac{\Delta T_{ei}(s)}{\Delta \omega_i(s)}$.

In order to arrive at these three distinct transfer functions, the machine speed and angle variation on the first transfer function must be held constant. This can be done by making the machine inertia very large. In the state space approach, an equivalent approach has been to treat speed $\Delta \omega_i$ of all machines as inputs by removing the rows of the state matrix A corresponding to the linearized differential equations $\frac{d\Delta \omega_i}{dt}$ and moving the columns corresponding to the speed state variable to append the input vectors. This provides useful information to examine the effect on the electrical torque $\Delta T_{ei}(s)$ of the ith machine as a result of perturbation in speed of the other machines ω_j. It also yields much required information on the interaction of two different machines at different frequency ranges. In the absence of the PSS, the transfer function $H_{\delta i}(s)$ through rotor angles provides a quantitative estimate of inherent damping. The imaginary component of this complex transfer function at a given frequency provides the requisite amount of damping. Generators having a negative or very small value of this damping are the key machines requiring additional contributions from PSS. The transfer function $H_\omega(s)$ provides information on the effectiveness of a PSS when connected in the system. This must have the desired shape that would counter the negative damping contribution through the rotor angle path transfer function $H_{\delta i}(s)$. A positive real part of $H_{\omega_{ij}}(s)$, i.e. the torque produced in the ith machine for speed perturbation at j^{th} machine, indicates that these two machines swing in phase and interactions between their PSS are not adverse as they jointly improve the overall damping. The parameters of the washout circuit, torsional filter etc. are tuned in an appropriate manner so as to have an ideal gain k_{pssi} over a range of modal frequencies, as described earlier. The magnitude of gain, on an individual machine base are then chosen. If after closed-loop eigenvalue analysis damping to the local and inter-area modes is not adequate the gains are adjusted. The transfer functions affecting the machine damping through various loops are characterized by the induced torque coefficient (ITC) subsequently to include the effect of FACTS device stabilizers (FDS) and for simultaneous co-ordination between PSS and FDS [Pourbeik and Gibbard, 1996, Pourbeik and Gibbard, 1998].

5.5.2 Frequency response approach

The frequency response approach has been used by many authors to design the PSS parameters. A number of references are available in [Martins, 2000]. The Nyquist technique was applied in [Gibbard, 1982] as a strategy to obtain the speed and power feedback based excitation controllers for a nominal set of gains. The concept of inverse Nyquist array (INA) was also applied to obtain the range of feedback gain in each feedback loop. The dynamic performance was tested. The procedure was repeated for other operating conditions and a gain was chosen from the range of gains. The primary objective was to introduce the phase characteristics of a PSS, modelled via lead-lag blocks $GPSS(s)$, to provide exact compensation for phase-lag introduced in the transfer function $GEP(s)$ i.e. pure damping at a complex frequency corresponding to the electromechanical modes. For a multi-machine system, the transfer function $GEP(s)$ has been denoted as $PVR(s)$ and methods to obtain $PVR(s)$ from state space descriptions of the system was also reported in [Kundur, 1994, Martins, 2000]. The parameters of the lead-lag blocks and and the gain to provide desired damping D_i were obtained by solving a set of non-linear equations:

$$GPSS_i(s) PVR_i(s) = D_i \tag{5.5}$$

$$\angle GPSS_i(s) + \angle PVR_i(s) = 0 \tag{5.6}$$

These three real non-linear equations were solved by Newton's method. Exact phase compensation, however, has never been provided in the field for the reasons discussed earlier. Kundur had demonstrated [Kundur et al., 1989] that from the perspective of overall dynamic performance of the system, phase must always be under-compensated and gain must be considered with regard to

Power System Stabilizers 69

the effect of noise in a practical situation. A theoretically optimal gain, however, could always be obtained from the root-locus plot. Gibbard [Gibbard, 1991, Gibbard and Vowles, 2004] showed that a fixed-parameter stabilizer, designed on the basis of phase compensation, would be robust to variation in operating conditions if a synthesized transfer function could be obtained to replace the several phase characteristics of the transfer function $PVR(s)$ for different operating conditions. Design and performance analysis has been carried out in a part of the Australian power network containing several machines.

5.5.3 Eigenvalue and state-space approach

Eigenvalue analysis is at the heart of the studies related to small signal stability. This approach is used extensively for tuning and analyzing the performance of PSS in a multi-machine system. The sensitivity of the electromechanical mode to the controller's parametric variations has been studied extensively [Pagola et al., 1989, Rouco and Pagola, 1997, Smed, 1993]. The eigenvalue sensitivity has served as a powerful tool for identifying the best generator locations [de Mello et al., 1980, Martins and Lima, 1990, Ostojic, 1988], calculating stabilizing signals [Martins and Lima, 1990], designing controller gain and quantifying interactions amongst multiple PSS [Martins, 2000]. The most appealing feature of eigenvalue sensitivity is that the conclusion is generic, as it is not limited by the complexities of the model.

Let us take a fixed structure controller $GPSS_{ij}(s) = k_{ij}H_{ij}(s)$ that would connect the jth output and i^{th} input of the plant $G_{ji}(s)$. We assume that the plant has a critical electromechanical mode λ_h whose damping is to be improved. The eigenvalue sensitivity i.e. changes in eigenvalue λ_h for a small changes in gain k_{ij}, can be expressed as [Pagola et al., 1989]

$$\frac{\partial \lambda_h}{\partial k_{ij}} = R_h^{ij} H_{ij}(\lambda_h) \qquad (5.7)$$

where R_{ij}^h is the residue of hth mode, i.e. the product of the modal controllability and modal observability, described earlier. The extension of (5.7) to a multi-input, multi-output (MIMO) system can be written as [Pagola et al., 1989]

$$\frac{\partial \lambda_h}{\partial k_{ij}} = trace\,[R_h H(\lambda_h)] \qquad (5.8)$$

For sufficiently small changes Δk_{ij} in gain k_{ij}, the shift of eigenvalue $\Delta \lambda_h$ is

$$\Delta \lambda_h^{ij} = R_h^{ij} H_{ij}(\lambda_h) \Delta k_{ij} \qquad (5.9)$$

We see that the phase of H_{ij} controls the direction of the shift of the mode λ_h and the changes in gain k_{ij} provides the amount of shift. The phase of

$H_{ij}(s)$ is selected to orient the direction of the mode towards the negative real axis in the complex plane. The mode can be moved more towards the left half plane by increasing the gain to achieve the desired closed-loop damping. Let us now interpret equation (5.9) for decentralized control structure, i.e. when the controller matrix is diagonal. Even though the output of one particular controller (say i^{th}) is driving one physical input point, this affects other outputs (besides i^{th}) because of the coupling in the generators through the network. The overall shift of a mode due to the change in the gain of the controller is the collective contribution of all the loops. The action of the change in gain k_{ii} is expressed as:

$$\Delta \lambda_h^{ii} = \sum_{j=1}^{n} R_h^{ji} H_{ii}(\lambda_h) \Delta k_{ii} \qquad (5.10)$$

The total shift of λ_h due to n controllers is

$$\Delta \lambda_h = \sum_{i=1}^{n} \sum_{j=1}^{n} R_h^{ji} H_{ii}(\lambda_h) \Delta k_{ii} \qquad (5.11)$$

The induced torque co-efficient (ITC) reported in [Ostojic, 1988, Gibbard, 1988, Pourbeik and Gibbard, 1998, Pourbeik and Gibbard, 1996] is equivalent to the elements R_h^{ji} in the residue matrix R_h.

A careful study of (5.10) is revealing. Usually, the phase of $H_{ii}(s)$ is selected to produce damping to mode h through G_{ii} with suitable gain k_{ii}. The phase and gain of the residue element R_{ii} helps in selecting the phase characteristics and the gain of the controller. The off-diagonal terms R_h^{ji}, even in a decentralized control structure, indicate to what extent different loops interact because of the coupling in the plant. Some of the coefficients may give rise to a rightwards movement of the mode. As a result, the net damping observed in the field might substantially differ from the design value. It is one of the primary sources of interactions amongst various control loops and the details are discussed in [Martins, 2000, Gibbard et al., 2000]. The effect of this interaction can be minimized by coordinated design of the PSS.

There are several methods of coordinated design that would produce the desired shift of λ_h leftwards. Methods based on optimization techniques are very effective for the design and coordination of a large number of PSS. The constrained linear programming approach has been applied successfully in [Doi and Abe, 1984, Pourbeik and Gibbard, 1998]. The problem is formulated as follows.

Let us assume that we need to coordinate the gain and the phase of n PSS units to improve m electromechanical modes $[\lambda_1, \lambda_2, ...\lambda_m]$. Let us also assume that the desired left shift $-\Delta\sigma = [-\Delta\sigma_1, -\Delta\sigma_2, ... - \Delta\sigma_m]$ and the change in frequency of the modes are within $\Delta\omega = [\Delta\omega_1, \Delta\omega_2...\Delta\omega_m]$. For m modes, (5.11) can be written in a compact form :

$$\Delta\lambda = \Phi k \qquad (5.12)$$

where

$$\Phi = \begin{bmatrix} \phi_{11} & \phi_{12} & .. & .. & \phi_{1n} \\ \phi_{21} & \phi_{22} & .. & .. & \phi_{2n} \\ .. & .. & .. & .. & .. \\ \phi_{m1} & \phi_{m2} & .. & .. & \phi_{mn} \end{bmatrix} \qquad (5.13)$$

$$\phi_{h,i} = \sum_{j=1}^{n} R_h^{ji} H_{ii}(h) \qquad (5.14)$$

$$k = [k_{11}, k_{22}, ...k_{nn}]^T \qquad (5.15)$$

A constrained linear programming problem is formulated as follows:

$$minimize \sum_{i=1}^{n} w_i \Delta k_{ii} \qquad (5.16)$$

sub :

$$Re(\Phi)\Delta K \leq -\Delta\sigma \qquad (5.17)$$

$$\Delta\mu \geq Im(\Phi)\Delta K \geq -\Delta\mu \qquad (5.18)$$

$$\Delta k_{max} \geq \Delta k \geq \Delta k_{min} \qquad (5.19)$$

An uniform weight w_i, say 1.0 for each, is usually set for all the controllers to treat their contribution equally. However, non-uniform weighting strategies have also been suggested. The limits on $\Delta\omega, \Delta k_{max}$ and Δk_{min} have to be set with care [Pourbeik and Gibbard, 1998]. Too high a value would introduce inaccuracy in the predicted mode shift as the relationship in (5.11) is based on small perturbations. It was feared that larger $\Delta\omega$ would interfere with first swing stability. On the other hand, too small a value was found to lead to infeasible solutions. One approach was to split the desired $\Delta\sigma$ into a several segments and repeat the optimization procedure for each one.

Pole-placement or eigenvalue assignments have been reported in the literature [Lefebre, 1983, Degtyarev and Cory, 1986, Lim and Elangovan, 1985, Elangovan and Lim, 1987, Chow and Sanchez-Gasca, 1989]. The basic idea of pole-placement for a SISO system has been described in [Kundur, 1994]. The primary disadvantage was that it was done without explicit requirements for robustness. The objective was to place a few selected eigenvalues at the desired locations. The problem has been formulated as an iterative optimization problem and solved through linear programming or with iterative approaches.

Linear quadratic control has been applied [Arnautovic and Medanic, 1987, Hopkins et al., 1981, Chow and Sanchez-Gasca, 1989] for coordinated control design. The problem has been formulated as a standard LQR and a full state feedback control was obtained from the solution of the relevant Riccati equation. An output feedback controller was then computed based on a subset of the eigen-space of the full state solution that retains the dominant modes of the closed-loop system. Structural constraints such as simple and decentralized control, feedback of only measured variables, etc. have been in use in power systems for many years and cannot be addressed by a standard LQR. Such a structurally constrained optimal control problem has been solved using the *Generalized Riccati Equation* [Geromel, 1987] and was applied to a power systems exploiting sparsity [Costa et al., 1997].

There has been substantial research to design a PSS employing methods that incorporated a requirement for an explicit robustness requirement in the design stage. Changing operating conditions have been modelled as uncertainties, structured or unstructured. The uncertainties have been modelled in the frequency domain through suitably defined weighting functions. In [Chow and Sanchez-Gasca, 1989], the desired response of the system in the post-fault steady state was characterized by a model and the action of the controller on the weighted plant was to minimize the error between the desired and actual behavior when subject to disturbances. This was termed *model matching control*. The $\mathcal{H}_\infty$ norm of the error was minimized by solving the relevant Riccati equations. Robust PSS design approaches, utilizing structural μ-synthesis techniques, have been reported in [Chen and Malik, 1995, Zhu et al., 2003]. The open-loop transfer function between the exogenous output to the disturbance was shaped by weighted filters prior to the control design. This is known as loop-shaping [Skogestad and Postlethwaite, 2001]. The uncertainties were considered to be structured as this gave a larger robustness margin. The optimization problem was solved by the *DK iteration technique* [Skogestad and Postlethwaite, 2001] available in μ-*synthesis toolbox* in Matlab. Optimization of the control effort for the desired level of damping has also been considered in [Soos and Malik, 2002]. In this approach, the PSS output was optimized by a suitably defined H_2 norm between the control output and the disturbance. One disadvantage of these analytical solution to these $\mathcal{H}_\infty$ norm optimization approaches was that the controller did not always guarantee adequate damping for lightly damped electromechanical modes. There has not been a direct correlation between minimization of the norm and the damping improvement of a particular mode. The success of these methods depended heavily on the judicious selection of weights for which no clear cut rule has been proposed to date.

If the restriction to an optimal control solution is relaxed, these $\mathcal{H}_\infty$ norm minimization approaches lead to the solution of a set of linear matrix inequalities (LMI)[Gahinet and Apkarian, 1994]. The interesting feature of LMI is that multiple objectives such as disturbance rejection via $\mathcal{H}_\infty$ control effort optimization through H_2 and pole-placement through a suitably defined LMI involving desired minimum damping ratio etc. can be be put together and solved as a multi-objective suboptimal control problem. The relevant constraints are expressed as a set of LMI through transformation. The problem has been solved by the interior point technique in convex programming [Nesterov and Nemirovskii, 1994]. The suboptimal control via the LMI approach has been applied for robust PSS design both employing state and output feedback [Rao and Sen, 2000, Werner et al., 2003]. The method has been particularly useful in power system damping design by offering a pole-placement objective in controller design specifications. A description of LMI is presented in later chapters where damping controllers employing FACTS devices are designed.

Recently evolutionary programming and intelligent control techniques such as genetic algorithms (GA), neural network and fuzzy logic have been applied to solve many complex optimization problems in engineering applications. With high speed computing tools, these search/rule based methods are increasingly being applied in power system planning, design, operation and control problems. The advantage of these methods is that the objective functions need not be explicit or differentiable and nonlinearity or non convexity are not a problem and minimal damping in the closed loop can be obtained.Constraints on controller performance, structure etc. can easily be imposed. The GAs are a set of heuristic search procedures based on the mechanics of natural selection, genetics and evolution [Goldberg, 1989]. The search task is in the parameter space including coding, fitness computation, population production, stopping criteria and the decoding of a binary stream into a physical parameter space. This technique has been applied to obtain damping ratios for electromechanical modes in a 10-machine 39-bus NETS system [Zhang and Coonick, 2000] and also in a model representing a portion of the Brazilian power network [do Bomfim et al., 2000].

The artificial neural network (ANN) technique produces the most likely outputs to specific inputs. It has been used extensively in load forecasting and economic despatch. A generic function is built from a series of past input-output relationships, an activity known as training. There are several algorithms that assist in the building of this trained model [Fausett, 1994] The accuracy of the ANN relies heavily on proper training. Several researchers, over the last decade, have demonstrated the merits of the ANN based PSS in different power system models [Zhang et al., 1993, Changaroon et al., 2000, Segal et al., 2000, Chaturvedi et al., 2004a, Chaturvedi et al., 2004b]. The primary objective of all these methods was to tune the parameters of the PSS where different schemes,algorithms and data sets for training and solutions were used.

Fuzzy set theory has been applied in many engineering disciplines [Mlynek and Patyra, 1996] including process control. In the fuzzy control approach, the inputs are transformed into a set of fuzzy variables through a process known as *fuzzification*. The entire range of input variable is split into several linguistic variables. Each linguistic variable is attached to a fuzzy membership function. Fuzzy variables are generated when these membership functions are applied to the precise values of the physical variables. The collection of such variables is known as an input fuzzy set. Each element in the input fuzzy set is mapped to an element in the output fuzzy set through fuzzy rules. The rules are defined by the designer based on the desired action of the controller. Results from off line simulations are often used. The knowledge of the operator and understanding of the dynamic behavior are usually employed to formulate the rules. The complexity of the rule sets grows with the number of linguistic variables and the number of inputs and outputs. More linguistic variables usually produce more accurate results. The fuzzy inputs through this rule based mapping produces inferences and these inferences fill in fuzzy output sets. The fuzzy output set is decoded to produce the desired control action. The merits of using fuzzy methods to PSS design have been shown by [Hosseinzadeh and Kalam, 1999, Hassan et al., 1991, El-Metwally et al., 1996, Hariri and Malik, 1996].

5.5.4 Summary

The function and action of PSS have been described.The Heffron-Phillips model has been used to show how the overall dynamic performance may be obtained.The merits of several feedback signals have been discussed.Guidelines for the choice of parameters-wash out time constants,phase and gain margins and PSS output limits- have been indicated. Various techniques of PSS design through damping torque, frequency response, eigenvalue and state-space approaches have been described. The application of optimal and linear robust control as an optimization problem are also discussed. Intelligent or heuristic approaches to robust tuning of PSS parameters were also briefly mentioned.

References

[Arnautovic and Medanic, 1987] Arnautovic, D. and Medanic, J. (1987). Design of decentralized multivariable excitation controller in multimachnie power systems by projective controls. *IEEE Transactions on Energy Conversion*, 2(4):598–604.

[Bayne et al., 1977] Bayne, J.P., Lee, D.C., and Watson, W. (1977). A power system stabiliser stabilising signal for thermal units based on derivation of accelerating power. *IEEE Transactions on Power Apparatus and Systems*, PAS-96(11):1777–1783.

[Changaroon et al., 2000] Changaroon, B., Srivastava, S. C., and Thukaram, D. (2000). A neural network based power system stabilizer suitable for on-line training-a practical case study for egat system. *IEEE Transactions on Energy Conversion*, 15(1):103–109.

[Chaturvedi et al., 2004a] Chaturvedi, D. K., Satsangi, P. S., and Kalra, P. K. (2004a). Experimental studies with a generalized neuron-based power system stabilizer. *IEEE Transactions on Power Systems*, 19(3):1445–1453.

[Chaturvedi et al., 2004b] Chaturvedi, D. K., Satsangi, P. S., and Kalra, P. K. (2004b). Performance of a generalized neuron-based pss in a multimachine power system. *IEEE Transactions on Energy Conversion*, 19(3):625–632.

[Chen and Malik, 1995] Chen, S. and Malik, O. P. (1995). Power system stabilizer design using μ synthesis. *IEEE Transactions on Energy Conversion*, 10(1):175–181.

[Chow and Sanchez-Gasca, 1989] Chow, J. H. and Sanchez-Gasca, J. J. (1989). Pole-placement designs of power system stabilizers. *IEEE Transactions on Power Systems*, 4(1):272–277.

[Costa et al., 1997] Costa, A. S., Freitas, F. D., and Silva, A. S. (1997). Design of decentralized controllers for large power systems considering sparsity. *IEEE Transactions on Power Systems*, 12(1):144–152.

[de Mello and Concordia, 1969] de Mello, F.P. and Concordia, Charles (1969). Concepts of synchronous machine stability as affected by excitation control. *IEEE Transactions on Power Apparatus and Systems*, 88(5):317–329.

[de Mello et al., 1978] de Mello, F.P., Mannett, L.N., and Undrill, J.M. (1978). Practical approaches to supplementary stabilising from accelerating power. *IEEE Transactions on Power Apparatus and Systems*, PAS-97(9):1515–1522.

[de Mello et al., 1980] de Mello, F.P., Nolan, P.J., Laskowski, T.F., and Undrill, J.M. (1980). Coordinated application of stabilizer for multi-machine power systems. *IEEE Transactions on Power Apparatus and Systems*, PAS-99(3):892–901.

[Degtyarev and Cory, 1986] Degtyarev, V.N. and Cory, B.J. (1986). Determination of stabilisers settings in multi-machine power system. *IEE Proc. C, Generation, Transmission and Distribution*, 133(6):308–313.

[do Bomfim et al., 2000] do Bomfim, A. L. B., Taranto, G. N., and Falcão, D. M. (2000). Simultaneous tuning of power system damping controllers using genetic algorithm. *IEEE Transactions on Power systems*, 15(1):163–169.

[Doi and Abe, 1984] Doi, A. and Abe, S. (1984). coordinated synthesis of power system stabilisers in multi-machine power system. *IEEE Transactions on Power Apparatus and Systems*, PAS-103:1473–1479.

[El-Metwally et al., 1996] El-Metwally, K.L., Malik, O.P., and Hope, G. S. (1996). Implementation of a fuzzy logic pss using a micro-controller and experimental test results. *IEEE Transactions on Energy Conversion*, 11(1):91–96.

[Elangovan and Lim, 1987] Elangovan, S. and Lim, C.M. (1987). Efficient pole-assignment method for designing power ssytem stabilizers for multi-machine power sytems. *IEE Proc. C, Generation, Transmission and Distribution*, 134(6):383–384.

[Fausett, 1994] Fausett, L. (1994). *Fundamentals of Neural Networks, Architecture, Algorithms and Applications*. Englewood Cliffs,Prentice-Hall,NJ.

[Gahinet and Apkarian, 1994] Gahinet, P. and Apkarian, P. (1994). A linear matrix inequality approach to H_∞ control. *International Journal of Robust and Non-linear Control*, 4(4):421–448.

[Geromel, 1987] Geromel, J.C. (1987). *Methods and Techniques for Decentralized Control Systems*. Clup, Italy.

[Gibbard, 1982] Gibbard, M.J. (1982). Coordination of multi-machine stabiliser gain settings for a specified level of system damping performance. *IEE Proc. C, Generation, Transmission and Distribution*, 129(2):45–48.

[Gibbard, 1988] Gibbard, M.J. (1988). Coordination design of multi-machine power system stabilisers based damping torque concepts. *IEE Proc. C, Generation, Transmission and Distribution*, 135(4):276–284.

[Gibbard, 1991] Gibbard, M.J. (1991). Robust design of fixed-parameter power system stabilisers over a wide range of operating conditions. *IEEE Transactions on Power Systems*, 6(2):794–800.

[Gibbard et al., 2000] Gibbard, M.J., Vowles, D. J., and Pourbeik, P. (2000). Interactions between, and effectiveness of, power system stabilizers and facts device stabilizers in multi-machine systems. *IEEE Transactions on Power Systems*, 15(2):748–755.

[Gibbard and Vowles, 2004] Gibbard, M.J. and Vowles, D.J. (2004). Reconciliation of methods of compensation for pss in multi-machine systems. *IEEE Transactions on Power Systems*, 19(1):463–472.

[Goldberg, 1989] Goldberg, D. E. (1989). *Genetic algorithms in search, optimization and machine learning*. Addision Wesley,Reading-MA.

[Gooi et al., 1981] Gooi, H.B., Hill, F.F., Mobarak, M.A., Thorn, D.H., and Lee, T.H. (1981). Coordinated multi-machine stabiliser setting without eigenvalue drift. *IEEE Transactions on Power Apparatus and Systems*, PAS-100(10):3879–3887.

[Hariri and Malik, 1996] Hariri, A. and Malik, O.P. (1996). A fuzzy logic based power system stabilizer with learning ability. *IEEE Transactions on Energy Conversion*, 11(4):721–727.

[Hassan et al., 1991] Hassan, M.A.M, Malik, O.P., and Hope, G. S. (1991). A fuzzy logic based stabilizer for a synchronous machine. *IEEE Transactions on Energy Conversion*, 6(3):407–413.

REFERENCES

[Heffron and Phillips, 1952] Heffron, W. G. and Phillips, R. A. (1952). Effect of modern amplidyne voltage regulators on underexcited operation of large turbine generators. *American Institutions of Electrical Engineers*, 71:692–697.

[Hopkins et al., 1981] Hopkins, W.E., Medanic, J., and Perkins, W. R. (1981). Output feedback pole placement in the design of suboptimal linear quadratic regulators. *International Journal of Control*, 34:593–612.

[Hosseinzadeh and Kalam, 1999] Hosseinzadeh, N. and Kalam, A. (1999). A direct adaptive fuzzy power system stabilizer. *IEEE Transactions on Energy Conversion*, 14(4):1564–1571.

[Kundur, 1994] Kundur, P. (1994). *Power System Stability and Control*. McGraw Hill, USA.

[Kundur et al., 1989] Kundur, P., Klein, M., Rogers, G.J., and Zywno, M.S. (1989). Application of power system stabilizers for enhancement of overall stability. *IEEE Transactions on Power Systems*, 4(2):614–626.

[Kuo and Golnaraghi, 2003] Kuo, B.C. and Golnaraghi, F. (2003). *Automatic Control Systems*. John Wiley and Sons, USA.

[Larsen and Swann, 1981] Larsen, E.V. and Swann, D.A. (1981). Applying power system stabilizers, part i, ii and iii. *IEEE Transactions on Power Apparatus and Systems*, PAS-100(6):3017–3046.

[Lee, 1992] Lee, D.C. (1992). *IEEE recommended practice for excitation system models for power system stability studies*. Energy development and power generation committee of power engineering society.

[Lee et al., 1981] Lee, D.C., Beaulieu, R.E., and Service, J.R.R. (1981). A power system stabilizer using speed and electrical power inputs -design and field experience. *IEEE Transactions on Power Apparatus and Systems*, PAS-100(9):4151–4167.

[Lefebre, 1983] Lefebre, S. (1983). Tuning of stabilisers in multi-machine power system. *IEEE Transactions on Power Apparatus and Systems*, PAS-102(1):290–299.

[Lim and Elangovan, 1985] Lim, C.M. and Elangovan, S. (1985). Design of stabilisers in multi-machine power system. *IEE Proc. C, Generation, Transmission and Distribution*, 132(3):146–153.

[Martins, 2000] Martins, N. (2000). Impact of interactions among power system controls. *CIGRE Special Publication 38.02.16*, Technical Brochure 166.

[Martins and Lima, 1990] Martins, N. and Lima, L.T.G. (1990). Determination of suitable locations for power system stabilizers and static var compensators for damping electromechanical oscillations in large power systems. *IEEE Transactions on Power Systems*, 5(4):1455–1469.

[Mlynek and Patyra, 1996] Mlynek, D.M. and Patyra, M.J. (1996). *Fuzzy Logic: implementation and applications*. John Wiley and Sons, UK.

[Nesterov and Nemirovskii, 1994] Nesterov, Y. and Nemirovskii, A. (1994). *Interior-point polynomial algorithms in convex programming*. Society for Industrial and Applied Mathematics, Philadelphia,USA.

[Ostojic, 1988] Ostojic, D. R. (1988). Identification of optimum site for power system stabiliser applications. *IEE Proc. C, Generation, Transmission and Distribution*, 135(5):416–419.

[Padiyar, 1996] Padiyar, K. R. (1996). *Power system dynamics : stability and control*. John Wiley; Interline Publishing, Singapore,Bangalore.

[Pagola et al., 1989] Pagola, F.L., Perez-Arriaga, I.J., and Verghese, G.C. (1989). On sensitivities, residues and participations: applications to oscillatory stability analysis and control. *Power Systems, IEEE Transactions on*, 4(1):278–285.

[Pourbeik and Gibbard, 1996] Pourbeik, P. and Gibbard, M.J. (1996). Damping and synchronising torques induced on generators by facts stabilisers in multimachine power systems. *IEEE Transactions on Power Systems*, 11(4):1920–1925.

[Pourbeik and Gibbard, 1998] Pourbeik, P. and Gibbard, M.J. (1998). Simultaneous coordination of power system stabilizers and facts device stabilizers in a multimachine power system for enhancing dynamic performance. *IEEE Transactions on Power Systems*, 13(2):473–479.

[Rao and Sen, 2000] Rao, P.S. and Sen, I. (2000). Robust pole placement stabilizer design using linear matrix inequalities. *IEEE Transactions on Power Systems*, 15(1):313–319.

[Rouco and Pagola, 1997] Rouco, L. and Pagola, F.L. (1997). An eigenvalue sensitivity approach to location and controller design of controllable series capacitors for damping power system oscillations. *IEEE Transactions on Power Systems*, 12(4):1660–1666.

[Segal et al., 2000] Segal, R., Kothari, M. L., and Madnani, S. (2000). Radial basis function (rbf) network adaptive power system stabilizer. *IEEE Transactions on Power Systems*, 15(2):722–727.

[Skogestad and Postlethwaite, 2001] Skogestad, S. and Postlethwaite, I. (2001). *Multivariable Feedback Control*. John Wiley and Sons, UK.

[Smed, 1993] Smed, T. (1993). Feasible eigenvalue sensitivity for large power systems. *IEEE Transactions on Power Systems*, 8(2):555–563.

[Soos and Malik, 2002] Soos, A. and Malik, O. P. (2002). An h_2 optimal adaptive power system stabilizer. *IEEE Transactions on Energy Conversion*, 17(1):143–149.

[Werner et al., 2003] Werner, H., Korba, P., and Yang, T. C. (2003). Robust tuning of power system stabilizers using lmi-techniques. *IEEE Transactions on Control Systems Technology*, 11(1):147–152.

[Zhang and Coonick, 2000] Zhang, P. and Coonick, A. H. (2000). Coordinated synthesis of pss parameters in multi-machine power systems using the method of inequalities applied to genetic algorithms. *IEEE Transactions on Power systems*, 15(2):811–816.

[Zhang et al., 1993] Zhang, Y., Chen, G. P., Malik, O. P., and Hope, G. S. (1993). An artificial neural network based adaptive power system stabilizer. *IEEE Transactions Energy Conversion*, 8(1):71–77.

[Zhu et al., 2003] Zhu, C., Khammash, M., Vittal, V., and Qiu, W. (2003). Robust power system stabilizer design using H_∞ loop shaping approach. *IEEE Transactions on Power Systems*, 18(2):810–818.

Chapter 6

MULTIPLE-MODEL ADAPTIVE CONTROL APPROACH

6.1 Introduction

A conventional damping control design approach considers a single operating condition of the system [Kundur, 1994]. A proportional-integral (PI) or a proportional-integral-derivative (PID) controller is designed to ensure desired performance under a particular operating condition. The controllers obtained from these approaches are simple but tend to lack robustness since, at times, they fail to produce adequate damping at other operating conditions. To address this issue, researchers, over the years, have proposed several adaptive control structures for power system stabilizers. Researchers [Malik et al., 1976] have applied the model reference adaptive control (MRAC) strategy where the error between the power system response and the reference model output is used to modify the controller parameters, such that the system behavior is driven to match the behavior of the reference model. A self tuning control (STC) of PSS has been reported in [Pahalawaththa et al., 1991] where the amount of pole shifting is adjusted depending upon the system conditions. Bandyopadhaya and Prabhu [Bandyopadhaya and Prabhu, 1988] have presented a gain scheduling control (GSC) scheme for PSS, where the controller parameters are tuned based on the minimization of the distance between the current and the desired operating points.

The primary concern in power system operation is that following a disturbance e.g. a fault on one of the buses, followed by outage of a part of the transmission network, the system switches to a different operating condition or network topology which are not known specifically in advance. From past statistics and study, one can have an approximate idea about the set of possible dynamics that are most likely to dictate the system behavior following such a disturbance however, the number of elements in this set might be very high. Moreover there exists a degree of uncertainty in the way the power system is likely to behave following a disturbance. Therefore, on-line identification is required to detect any trend in the post-disturbance dynamic behavior and switch an appropriately weighted combination of pre-tuned controllers.

One such adaptive algorithm is the multiple-model adaptive control (MMAC), which was originally introduced by Lainiotis [Lainiotis, 1976]. Subsequently, it has been used for the control of aircraft [Athans et al., 1977] and for regulation of hemodynamic variables [He et al., 1986, Martin et al., 1987]. The concept was extended for power system damping control design in [Chaudhuri et al., 2004]. Our basic motivation behind applying this scheme in power system model is that it can achieve the desired performance without any requirement to identify the post-disturbance dynamics prior to initiating the control action. The assumption, though, is that the actual system response can be represented by a single or a suitable combination of a finite number of linearized models. Separate controllers (PI,PID) are also assumed to be designed a priori to ensure satisfactory performance for each of these models.

Theoretically, one can not claim that a convex combination of stabilizing controllers necessarily produce a stable closed-loop response. However, it has been found that the MMAC strategy has produced adequate stability margin and robustness for a range of test cases considered in our research.

6.2 Overview of MMAC strategy

A schematic overview of the conventional multiple-model adaptive control (MMAC) scheme is given in Fig. 6.1.

The recursive algorithm uses a bank of linearized system models to capture the possible system dynamics following a disturbance. One separate controller k is designed and tuned, a priori, based on each model k from the model bank. At each simulation step, the actual system response is compared with the response of the linearized models which are driven by the same control input. The differences in the response of each model with respect to the actual system response is used to generate individual model residuals. Using these residuals, the probability of each model representing the actual system response is computed. Based on the probabilities, suitable weights are assigned to individual control moves such that the less probable models carry less weight. This ensures that the controllers designed for the less probable models influence the final control move to a lesser extent. The resultant control action is, thus, a probability weighted average of the control moves of each individual controller.

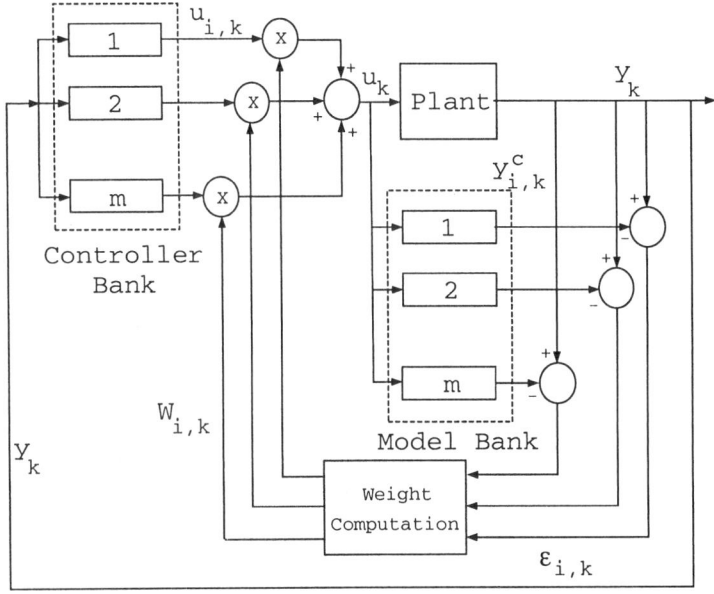

Figure 6.1. Schematic of MMAC strategy

At each stage of the recursive algorithm, primarily two tasks are performed i.e. calculation of probability using a Bayesian approach and assignment of suitable weights based on the probability value.

6.2.1 Calculation of probability: Bayesian approach

We have used the recursive Bayes theorem for computing the probability of each model in the bank. The theorem calculates the conditional probability of the ith model in the model bank being the true model of the system given this model population. The probabilities are assumed to be stochastic and Gaussian in nature and thus take a form of the exponential of the negative square of the residuals [Rao et al., 2003]. At the kth step, the probability for the ith model is calculated as:

$$p_{i,k} = \frac{\exp\left(-\frac{1}{2}\varepsilon_{i,k}^T C_f \varepsilon_{i,k}\right) p_{i,k-1}}{\sum_{j=1}^{N} \exp\left(-\frac{1}{2}\varepsilon_{j,k}^T C_f \varepsilon_{j,k}\right) p_{j,k-1}} \quad (6.1)$$

where,

$$\varepsilon_{i,k} = y_k - y_{i,k}^c \quad (6.2)$$

is the error or model residual at the current step. N denotes the total number of models in the model bank and C_f is the convergence factor that is used to tune the rate of convergence of the probabilities. Large values of C_f will magnify the model residuals and cause an acceleration of convergence to a single model. The recursion is initialized by assigning equal probability ($1/N$) to all the models in the bank. At each iteration, new probabilities are calculated thereby improving upon the probability computed at the previous iteration. One major advantage is that this algorithm is computationally inexpensive. An additional benefit is that the poor models are rejected exponentially and thereby allowing to a widely varying set of models without necessarily leading to a large drop in controller performance, even during the initial stages [Yu et al., 1992].

To summarize, for a given set of models, the above algorithm recursively determines the probability that the ith model is the true system model. The computation is based on the present model residuals with respect to the actual system response and the previous probabilities for each model [Rao et al., 2003].

6.2.2 Calculation of weights

Based on the probability of individual models, calculated during each recursive step, suitable weights are assigned to the control actions of each of the controllers. The model with a higher probability is assigned a higher weight and vice versa. One of the feature of this Bayesian approach is that it can only assume a steady-state probability of either zero or one and consequently, the algorithm converges to a single model. However, due to the uncertainties associated with a practical power system, it is unlikely that any single model in the model bank would be exactly equivalent to the system under control, and hence proper blending of control action is often required. Models attaining a probability of zero cannot enter the subsequent recursions and hence an artificial cut-off β_{min} is used to keep them alive. At the kth step, the ith model is assigned a weight $W_{i,k}$ such that:

$$W_{i,k} = \begin{cases} \frac{p_{i,k}}{\sum_{j=1}^{N} p_{j,k}} & \forall p_{i,k} > \beta_{min} \\ W_{i,k} = 0 & \forall p_{i,k} < \beta_{min} \end{cases} \quad (6.3)$$

For models with $p_{i,k} < \beta_{min}$, the probability is reset to $p_{i,k} = \beta_{min}$ and these models are then excluded from being weighted. At the kth iteration, the resulting probability-weighted control move is computed as:

$$u_k = \sum_{j=1}^{N} W_{j,k} \cdot u_{j,k} \quad (6.4)$$

6.3 Study system

The MMAC scheme involves a number of linearized system models corresponding to different operating conditions. To illustrate the principle, a simple study system is considered to start with and later the method is demonstrated using the study system described in Chapter 4. A simple 4-machine, 2-area study system is shown in Fig. 6.2.

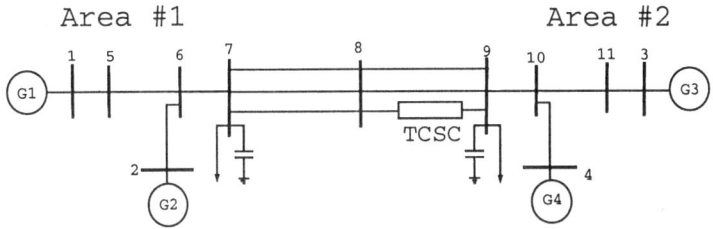

Figure 6.2. Study system

This system is considered as one of the benchmark models for performing studies on inter-area oscillation because of its realistic structure and availability of system parameters [M.Klein et al., 1991, Kundur, 1994]. All four generators are represented using the sub-transient model with DC (IEEE-DC1A type) excitation system as described in Chapter 4. The detailed dynamic data for the system can be found in [Kundur, 1994]. The system consists of two areas connected by a weak transmission corridor. To enhance the transfer capability of the corridor, a TCSC is installed in one of the lines connecting buses #8 and #9 as shown in Fig. 6.2. From the transfer capacity enhancement point of view, the percentage compensation k_c of the TCSC is set to 10%. A maximum and minimum limit of 50% and 1%, respectively, is imposed on the dynamic variation of k_c. Under normal operating conditions, the power flow from area #1 to area #2 is 400 MW. Eigen value analysis for this base case, displayed in Table 6.1, shows the presence of one lightly damped inter-area mode and two reasonably damped local modes of oscillation [M.Klein et al., 1991].

Table 6.1. Critical modes of oscillation of the study system

Mode	ζ	$f(Hz)$
Inter-area	0.0129	0.6308
Local	0.0809	1.0813
Local	0.0789	1.1159

Due to the lightly damped mode, there would be inter-area oscillations following a disturbance in the system. The objective, therefore, is to design a damping control scheme for the TCSC to mitigate these unwanted oscillations. Moreover, the control action should be robust with respect to varying operating conditions. The real power flow in the line connecting buses #10 and #9 was chosen as the feedback stabilizing signal for the controller since the inter-area mode was found to be highly observable [Martins and Lima, 1990] in this measured signal.

6.4 Model bank
6.4.1 4-machine, 2-area system

More than ten linearized small-signal models were required to span the entire space of anticipated response of the system following a disturbance. Disturbances include either a bus fault rendering outage of a line or a sudden change in power flow through the key tie-lines or a change in the nature of the loads etc. Corresponding to each of the post-disturbance operating conditions, different linearized models of the system were obtained. Ideally, each one of them should have been included in the model bank. However, to reduce computation time, only the five most probable models, in terms of their likelihood to represent the actual system response, were used. The operating scenarios and corresponding model identifiers are summarized in Table 6.2.

Table 6.2. Operating conditions used in the model bank

Model No.	Tie-line flow (MW)	Outage of line
1	400	no outage
2	400	7 - 8
3	400	8 - 9
4	300	no outage
5	500	no outage

Model #1 is for the nominal operating condition with 400 MW power transfer through the corridor and all the tie-lines in place. Model #2 reflects the situation with one of the tie-lines between buses #7 and #8 switched off. Model #3 corresponds to an outage of one of the tie-lines connecting buses #8 and #9. In both the above cases, the tie-line power flow were assumed to remain unchanged at 400 MW. Two different tie-line power flows of 300 MW and 500 MW between area #1 and area #2 were represented by models #4 and #5, respectively with all the tie-lines in opeartion.

6.4.2 16-machine, 5-area system

This study system is described in Chapter 4. Only a few credible contingencies are considered for building the model bank for a relatively large power system model like this one. For example a fault in the two main transmission corridors between bus #53-#54 and #60-#61 are severe contingencies. Also the sudden change in power flow through a line has been included. In this way nine probable system models have been considered for which the operating scenarios and corresponding model identifiers are summarized in Table 6.3.

Table 6.3. Operating conditions used in the model bank

Model	Tie-Line flow (MW)	Outage of line	Type of load
1	700	No outage	CI
2	700	53-54	CI
3	700	60-61	CI
4	700	27-53	CI
5	100	No outage	CI
6	900	No outage	CI
7	700	No outage	CI + CC
8	700	No outage	CI + CP
9	700	No outage	Dynamic load

Model #1 is for the nominal operating condition with 700 MW power transfer through the tie-lines and all the tie-lines in place. Model #2, #3 and #4 reflect the situation when one of the tie-lines between buses #53-#54, #60-#61 and #27-#53 are switched off respectively. Model #5 and #6 correspond to 100 MW and 900 MW through tie line 60-61 respectively. Model #7 and #8 are combinations of constant current (CC), constant power (CP) and constant impedance (CI) load. In model #9, a dynamic induction motor type load is considered at bus #41.

6.5 Control tuning and robustness testing
6.5.1 4-machine, 2-area system

The first step towards the implementation of the MMAC scheme is to design and tune the controllers for the five linearized system models, described in Table 6.2. The order for each of these system models was 41 but to facilitate control design, they were reduced to 3rd order equivalents. The simplified system frequency response is shown in Fig. 6.3 with respect to the original one for the nominal operating condition.

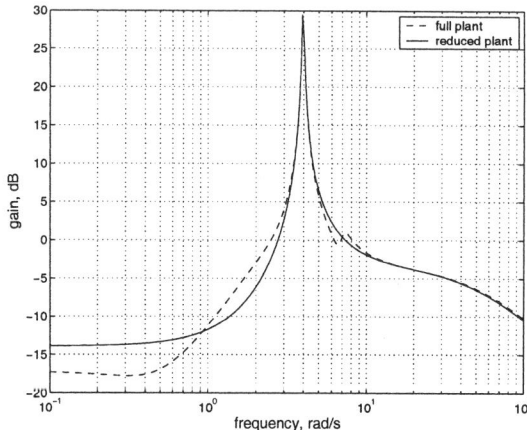

Figure 6.3. Frequency response of original and simplified plant

To improve the damping ratio of the critical inter-area mode, a simple controller, as shown in Fig. 6.4, was designed for the reduced system model using the conventional gain-margin and phase-margin based techniques [Kuo and Golnaraghi, 2003]. The controller gain required additional tuning to meet the specified closed-loop performance criteria. In this case, the criterion was to achieve a closed-loop damping ratio of 0.25 for the inter-area mode under all operating conditions. Our experience is that a damping ratio of 0.25 generally ensures satisfactory settling of inter-area oscillations within 10-12 s, a criterion followed by the power system utilities [Paserba, 1996].

The controller gains were adjusted individually for each model, using root locus techniques to achieve a damping ratio of 0.25 for the inter-area mode i.e. the controller k was tuned so as to ensure a closed-loop inter-area mode damping ratio of 0.25 for model k. However, this did not necessarily ensure that satisfactory damping ratios would be preserved using controller k for plant models other than k. In fact, it is clear from Table 6.4 that in certain cases, either the system becomes unstable or the damping ratio is below the acceptable limit. For the cases marked as 'unstable' in Table 6.4, the damping ratio for the inter-area mode was acceptable, but some of the other modes had negative damping ratios.

If the controllers were tuned to obtain a less conservative damping ratio of 0.15 instead of 0.25, then the instabilities could be avoided in some cases, but some of the damping ratios under certain operating conditions went below 0.1, which is not acceptable for secure operation of the power system. It is to be noted that although the above discussion is specific to this particular test

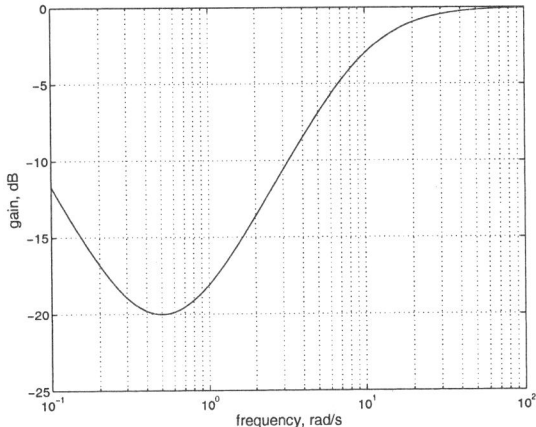

Figure 6.4. Frequency response of the controller

Table 6.4. Closed-loop damping ratio of the inter-area mode for different models and controllers

Controller No.	Model No.				
	1	2	3	4	5
1	0.25	0.23	unstable	0.17	unstable
2	0.26	0.25	unstable	0.18	unstable
3	0.16	0.15	0.25	0.11	0.22
4	unstable	unstable	unstable	0.25	unstable
5	0.18	0.17	0.27	0.12	0.25

system, it still represents the general lack of robustness of the conventionally tuned controllers under different operating conditions encountered in a practical power system.

6.5.2 16-machine, 5-area system

To improve the damping ratio of the critical inter-area modes, an observer based state feedback controller was designed for each of the models. The state feedback gain was determined to ensure settling of inter-area oscillations within 10-12 s. The 'place' function available with the *Control System Toolbox* in *Matlab* was used to compute the gain. In power systems, all the states are not always available for feedback and hence an observer was designed to derive these states from the measured outputs.

However, controllers designed using such a technique cannot maintain their desired performance level for a range of possible operating conditions. We have examined the performance of the designed controllers against different post-disturbance conditions described by the linearized models in the model bank. A few simulation results for plant model and controller combinations are shown in Figs. 6.5 and 6.6.

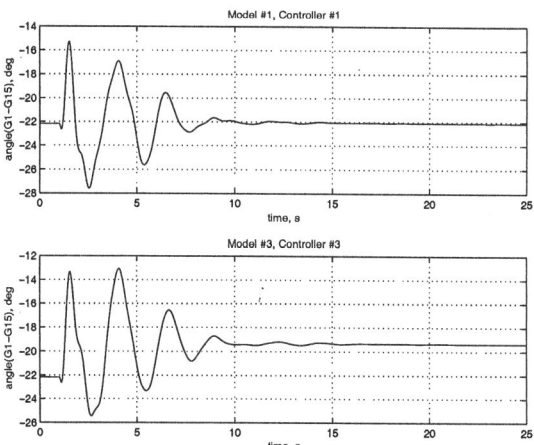

Figure 6.5. Performance of conventional controllers

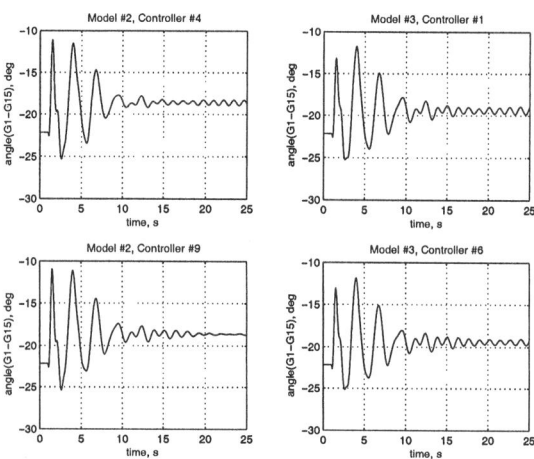

Figure 6.6. Robustness test for conventional controllers

It can be seen in Fig. 6.5 that model #1 - controller #1 combination have well settled response, as expected. Similar satisfactory performances were obtained

Multiple-model Adaptive Control Approach 89

for models with the corresponding controller. As an example the performance of model #3 is shown with controller #3. However, the responses for model #2 - controller #4, model #2 - controller #9, model #3 - controller #1 and model #3 - controller #6, shown in Fig. 6.6, are inferior as oscillations of a small magnitude continue beyond 15 s. This has provided the motivation to adopt a probabilistic approach for improving the robustness under different operating conditions.

6.6 Test cases

It is clear from the results shown in Table 6.4 and Fig. 6.6 that a conventional controller k, designed and tuned on the basis of model k, is not necessarily guaranteed to meet the desired performance specification for other models. Therefore, some mechanism needs to be devised for on-line identification of the unknown dominant dynamics following a disturbance and switch to an appropriately weighted combination of the controllers. Two situations can arise considering the uncertainty involved in a practical power system and the limit on the number of models that can be included in the model bank from the computational complexity point of view. In one case, the model corresponding to the dominant post-disturbance dynamics is likely to be present in the model bank, hence, the scheme should pick up the controller corresponding to that model with maximum weight. In the other case, the model representing the dominant post-disturbance dynamics is less likely to be present in the model bank, so, the scheme should be able to ensure proper blending between the control moves of the existing controllers to achieve the desired performance criteria. These two test cases have been treated separately in this chapter and are elaborated in the following subsections.

6.6.1 Test case I

For the 4-machine, 2-area system (Fig. 6.2), a solid three phase line to ground fault was simulated at bus #8 for 80 ms, followed by opening of one of the tie-lines connecting buses #7 and #8. From Table 6.2, it can be seen that the dynamics corresponding to this particular post-disturbance situation is captured in model #2. All five models, including model #2, were kept in the model bank and the corresponding controllers in the controller bank. The objective was to see whether and how quickly the adopted MMAC algorithm could identify the dominant post-disturbance dynamics and switch to the appropriate controller to achieve the desired performance.

For the 16-machine, 5-area system (Fig. 4.1), a three phase solid line to ground fault was simulated at bus #53 for 80 ms, followed by opening of one of the tie-lines connecting buses #53 and #27. From Table 6.3, it can be seen that the dynamics corresponding to this particular post-disturbance situation is captured by model #4. All nine models, including model #4, were kept in the model bank and the corresponding controllers in the controller bank.

6.6.2 Test case II

The same disturbance, as described before, was considered. Because of the uncertainty involved in a practical power system, it is unlikely that any single model in the model bank would be the exact equivalent of the system under control. Moreover, due to computational constraints, only a few out of the large number of possible models can be included in the model bank. To replicate these two likely situations, model #2 and the corresponding controller #2 were deliberately removed from the respective banks. The idea was to validate whether a blended version of the remaining control moves is able to achieve the desired performance in the absence of the actual controller. This demonstrates the ability of the MMAC algorithm to pick up a proper blend of the relevant post-disturbance dynamics to closely mimic the actual system response.

For the 16-machine, 5-area system, two simulation studies were done. In the first case, the same disturbance, as described in Test case I, was considered. In the second case, a sudden partial loss in generation at generator G1 has been imposed. Model #4 and the corresponding controller #4 was deliberately removed from the banks for the first simulation study. The model corresponding to the second disturbance was absent in both the model and controller bank. The idea, as before, was to determine whether an appropriate blending of the remaining control moves was able to achieve the desired performance in the absence of the actual controller.

6.7 Choice of convergence factor and artificial cut-off

Two of the most important factors influencing the success of a MMAC scheme are the proper choice of the convergence factor (C_f) and the artificial cut-off (β_{min}), described in (6.1) and (6.3) respectively. The choice, of course, is very much dependent on the specific system to be controlled and the design of the model banks. Although there are no hard and fast rules for choosing these parameters, a general guideline can be presented. Fig. 6.7 shows the time variation of the computed weights for some selected values of C_f and β_{min}.

It can be seen that with increasing value of C_f, the poor models are rejected quickly, whereas lower values of C_f help the blending. Higher values of the cut-off β_{min}, on the other hand, retain even the least probable models to help this blending. If there is a high chance that the post-disturbance behavior would be dominated by one of the models in the model bank, then it is preferable to use a high value of C_f to quickly reject the unwanted models and a low value

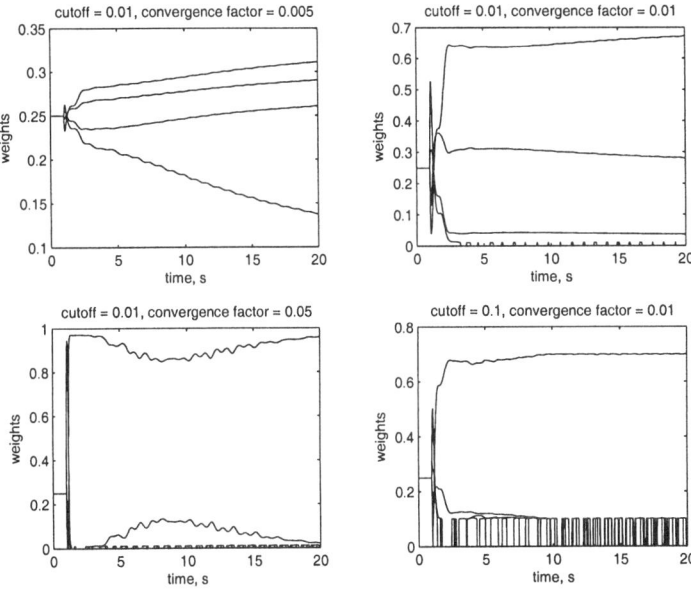

Figure 6.7. Variation of the computed weights

of β_{min} to prevent them from being retained during recursion. For a practical power system, this might not always be the relevant scenario. In practice, the number of probable models is too large for them all to be included in the model bank keeping the computational constraints in mind. Moreover, due to the uncertainties involved in the parameters, it is unlikely that any single model in the model bank would be exactly equivalent to the system under control. The calculated values of model residuals during the initial stages might be misleading in the sense that the dynamics of the system during the fault are often completely different from those during the post-fault situation. Therefore, instead of quickly rejecting the majority of the models based on the initial model residuals, blending is preferred by using lower values of C_f and higher values of β_{min}.

6.8 Simulation results with a 4-machine, 2-area study system

Simulations were performed in the *Simulink* [sim, 2002] environment of *Matlab* using a fixed step-size of 1 ms and 4th order Runge-Kutta solver. The results are separately presented for the two test cases.

6.8.1 Test case I

The results of the time domain simulation for Test case I are shown in Figs. 6.8 to 6.11. Here, the linearized model of the power system corresponding to the post-disturbance situation (model #2) was considered to be present in the model bank. As a result, the residual for model #2 starts decreasing after a few initial recursive steps and consequently the weight corresponding to this model goes up to attain a steady-state value of almost 1.0, see Fig. 6.8.

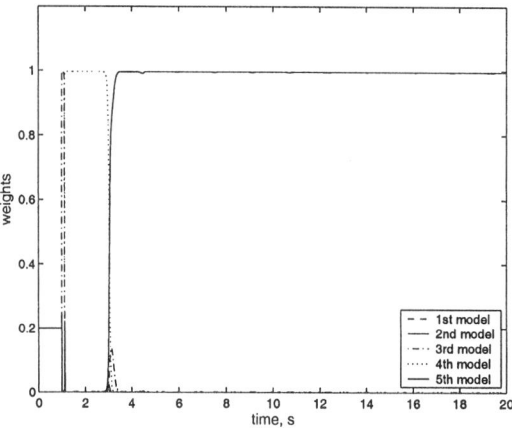

Figure 6.8. Test case I : Variation of the weights corresponding to each model

Our objective, in this case, is to demonstrate the ability of the MMAC scheme to identify the unknown dynamics and switch to the proper controller. This is why a relatively high value 0.05 was chosen for the convergence factor C_f to quickly reject the unwanted models. Also, the artificial cut-off β_{min} is kept to a small value of 0.001 to avoid retaining these unwanted models during subsequent recursive steps. If more blending is desired, both β_{min} and C_f can be adjusted accordingly as illustrated in the previous section.

The dynamic behavior of the system in response to the disturbance described previously is depicted in Fig. 6.9. The displays show the relative angular separation between machines #G1, #G4 and #G3, #G2. Inter-area oscillation involves a group of machines in one area swinging against a group in another area and is, therefore, mostly manifested in these particular relative angular differences. It can be seen that the lightly damped oscillations are settled in 10-12 s in the presence of the applied control scheme. Power flow in the line connecting buses #10 and #9, shown in Fig. 6.10, also settles within the stipulated time-frame. The sharp fall in the power flow, just after 1.0 s, is due to the inception of the fault which is cleared after 80 ms. Fig. 6.11 shows the resultant control ac-

Multiple-model Adaptive Control Approach 93

tion, which is dominated by the response of controller #2, because of its higher weight as shown in Fig. 6.8.

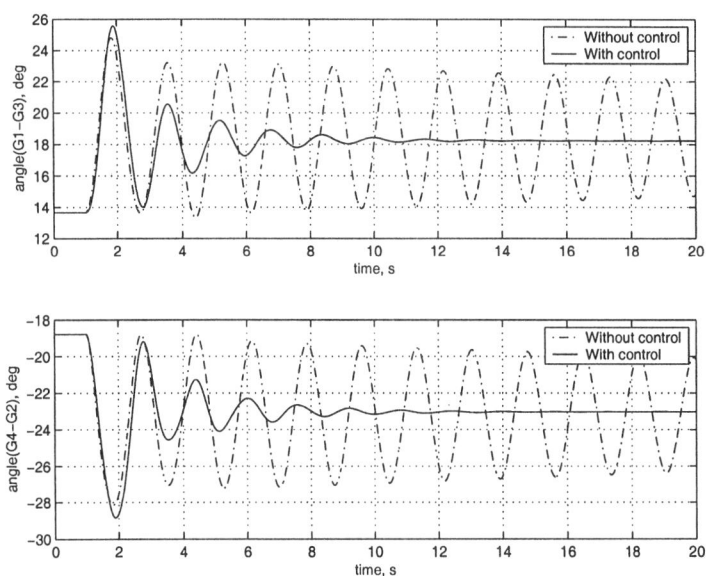

Figure 6.9. Test case I : Dynamic response of the system

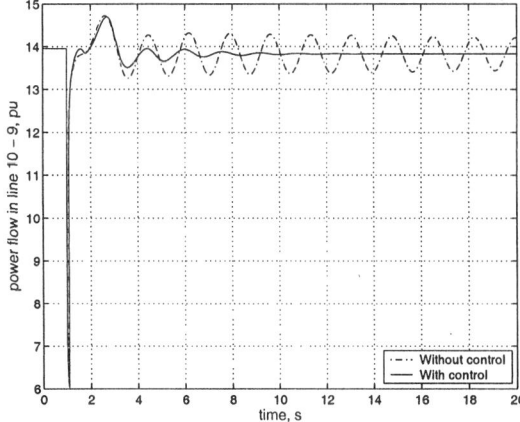

Figure 6.10. Test case I : Power flow between buses #10 and #9

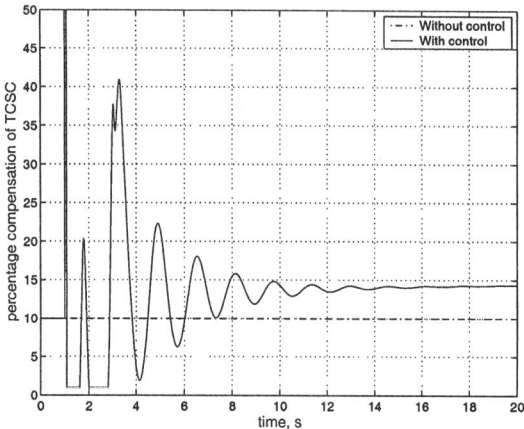

Figure 6.11. Test case I : Response of the controller

Simulation results illustrate that the MMAC scheme is able to identify the predominant post-disturbance dynamics and switch in the appropriate controller without any prior knowledge about the specific operating condition. The weights corresponding to the other pre-designed controllers decay exponentially to the minimum cut-off level. This ensures that the system performance is nearly close to the optimum considering that the post-disturbance dynamic behavior of the system is quite likely to be governed by one of the models in the bank.

6.8.2 Test case II

The results of the time domain simulation for Test case II are shown in Figs. 6.12 to 6.15. Contrary to the previous case, the linearized model (model #2) of the power system governing the post-disturbance dynamics and the corresponding controller (controller #2), was intentionally removed from the model bank. As a result, none of the model weights attains a steady-state value of almost 1.0, unlike the previous case, see Fig. 6.12.

After a few recursive steps, during which the trend is not very clear, it can be seen that the dynamics are governed primarily by models #4, #1 and #5, in that order. As before, the amount of blending can be adjusted by changing C_f, and/or β_{min}. In this case, the value of C_f was chosen to be relatively low (0.01) as the chances of converging to a single model is less. Also, the magnitude of the artificial cut-off β_{min} was increased to 0.01 to retain even the least probable models.

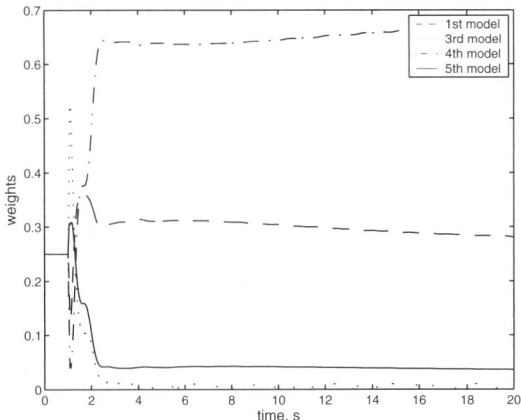

Figure 6.12. Test case II : Variation of the weights corresponding to each model

Fig. 6.13 exhibits the dynamic behavior of the system in response to the same disturbance as in the previous case. It can be seen that the lightly damped inter-area oscillations are settled in 10-12 s. Power flow between buses #10 and #9, shown in Fig. 6.14, also settles within the specified time. Fig. 6.15 shows the resultant control action, which is dominated by the response of controllers #4, #1 and #5 due to their relatively higher weights, as shown in Fig. 6.12.

The simulation results illustrate that, even though the actual model governing the response of the system after the disturbance is absent, the MMAC scheme is able to properly blend the control moves of the remaining controllers and still maintain reasonably similar performance. In fact, no noticeable deterioration can be observed in terms of performance in Fig. 6.13, when compared with Fig. 6.9. During the fault, the dynamics of the system is represented in a more realistic way by a combination of several models rather than a single model. This is particularly encouraging as it makes the MMAC scheme a reasonable candidate for application in large practical power systems, where the chances of convergence to a single model are remote, as described earlier.

6.9 Simulation results with a 16-machine, 5-area study system

6.9.1 Test case I

The results of time domain simulation for Test Case I are shown in Figs. 6.16 and 6.17.

The linearized model of the power system corresponding to the post-disturbance situation (model #4) was present in the model bank. As a result, the error residual for model #4 starts decreasing after a few initial recursive steps and

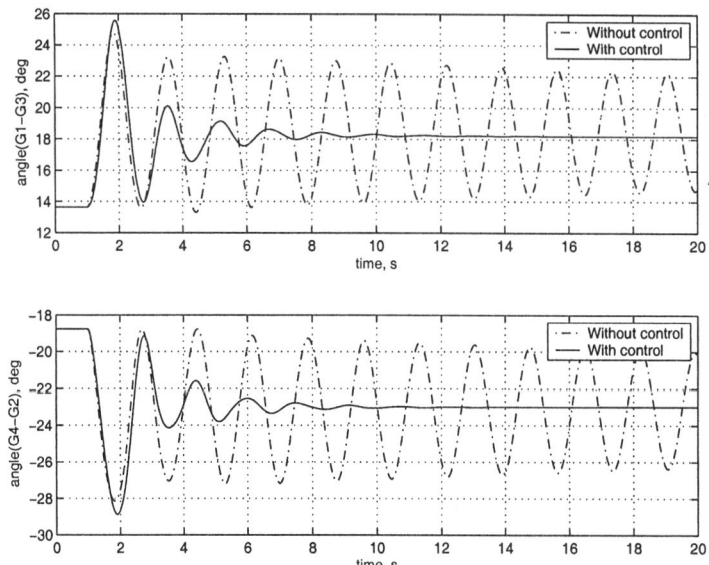

Figure 6.13. Test case II : Dynamic response of the system

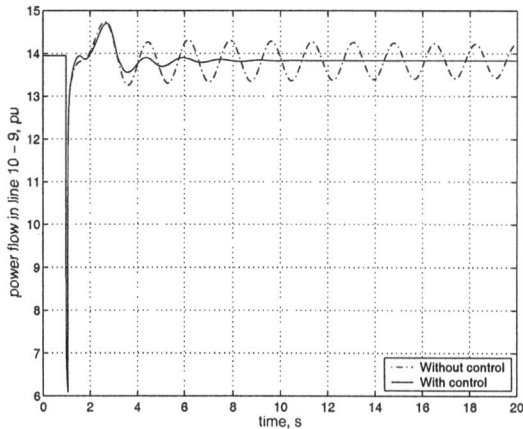

Figure 6.14. Test case II : Power flow between buses #10 and #9

consequently the weight corresponding to this model increases and attains a steady-state value of almost 1.0, see Fig. 6.16. Our aim, in this case, was to demonstrate the ability of the control scheme to identify the unknown dynamics and switch to the proper controller. This is why a relatively high value 0.005 was chosen for the convergence factor C_f to quickly reject the unwanted mod-

Multiple-model Adaptive Control Approach

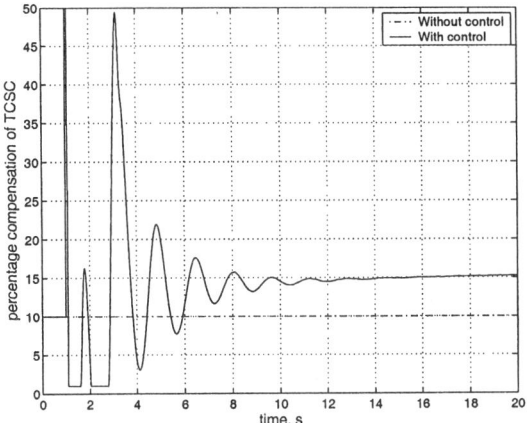

Figure 6.15. Test case II : Response of the controller

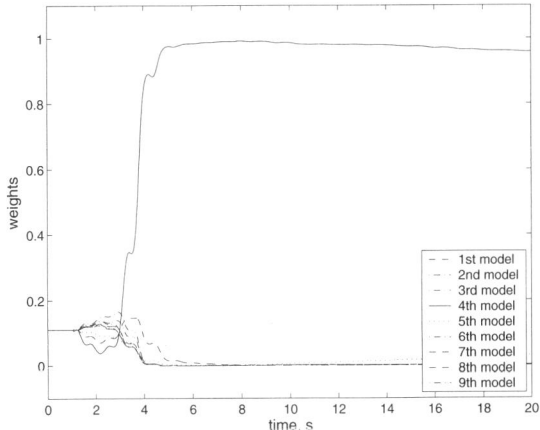

Figure 6.16. Test Case I: Variation of weights

els. Also, the artificial cut-off β_{min} was kept to a small value of 0.0001 to avoid retaining these unwanted models during subsequent recursive steps.

The displays in Fig. 6.17 show the relative angular separation between machines G1-G15 and G14-G13. It can be seen that the lightly damped oscillations are settled in 12-15 s in the presence of the applied control scheme. Power flow between buses #60 and #61 also settles within the acceptable time frame. The resulting control is primarily dominated by controller #4 owing to its higher weight. The simulation results illustrate that the control scheme is able to identify the dominant post-disturbance dynamics and switch to the

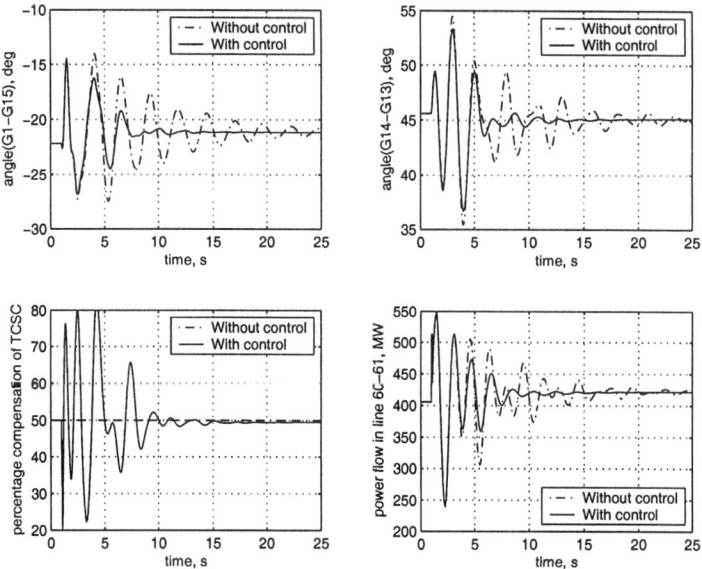

Figure 6.17. Test Case I: Dynamic response of the system

proper controller without any prior knowledge about the post-disturbance operating condition by using on-line recursive calculation of model probabilities and associated weights.

6.9.2 Test case IIa

For these two simulation studies, the model governing the post-disturbance dynamics were not present in the model bank and also the corresponding controllers were absent from the controller bank. The simulation results for Test Case IIa are shown in Figs. 6.18 and 6.19. Contrary to the previous case, the linearized model (model #4) of the power system governing the post-disturbance dynamics and the corresponding controller (controller #4), was intentionally removed from the model bank. As a result, weights corresponding to none of the models attain a steady state value of almost 1.0, unlike the previous case, see Fig. 6.18.

As before, the amount of blending can be adjusted by changing C_f, and/or β_{min}. In this case, the value of C_f was chosen to be relatively low (0.0001) as the chances of converging to a single model is less. Also, the magnitude of the artificial cut-off β_{min} was increased to 0.01 to retain even the least probable models. Fig. 6.19 exhibits the dynamic behavior of the system in response to the same disturbance as in the previous case.

Multiple-model Adaptive Control Approach

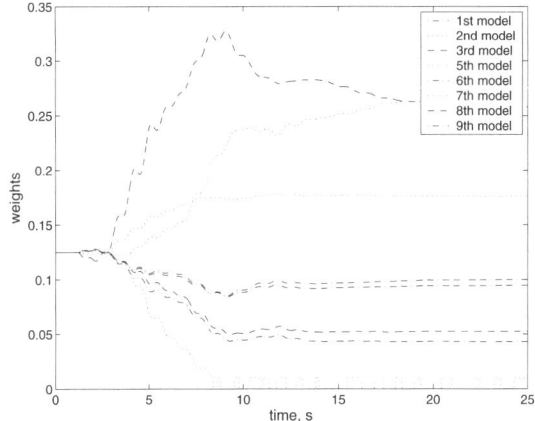

Figure 6.18. Test Case IIa: Variation of weights

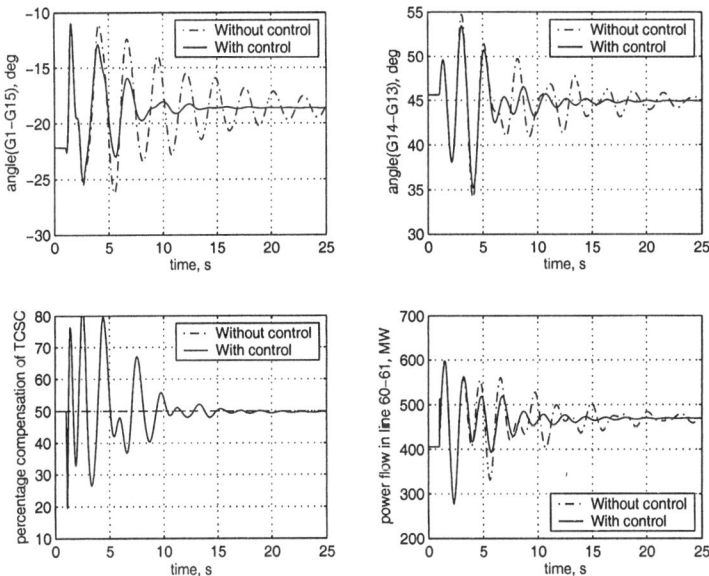

Figure 6.19. Test Case IIa: Dynamic response of the system

It can be seen that the lightly damped inter-area oscillations are settled in 12-15 s. Power flow between buses #60 and #61 also settles within the specified time. The simulation results illustrate that, even though the actual model governing the response of the system after the disturbance is absent, the control scheme is able to properly blend the control moves of the remaining controllers and still maintain reasonably similar performance. It should be noted that the compensation of the TCSC varies from 20% to 80% in both the cases as shown in Figs. 6.19 and 6.17. The range of variation is relatively large as a single centralized controller is used to damp all three inter-area modes. This is one of the drawbacks of a centralized controller.

6.9.3 Test case IIb

The simulation results for Test case IIb are shown in Figs. 6.20 and 6.21.

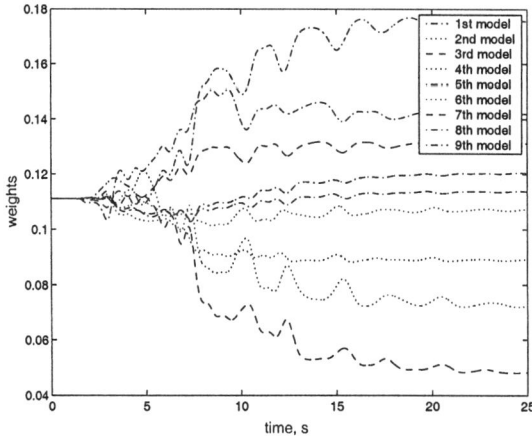

Figure 6.20. Test Case IIb: Variation of weights

In this case, a sudden partial generation loss was considered at generator G1. After 1 s, generation at generator G1 was reduced to 25% of its rated capacity, a contingency not considered while building the linearized models and corresponding controllers. The same values of C_f and β_{min} were used as in Test case IIa to ensure proper blending. The system response is shown in Fig. 6.21.

The simulation results illustrate that, even though the actual model governing the response of the system after the disturbance is absent, the control scheme is able to properly blend the control moves of the remaining controllers and still maintain reasonably similar performance. In fact, no noticeable deterioration can be observed in terms of performance in Fig. 6.17, when compared with Figs. 6.19 and 6.21. This is particularly encouraging as it makes this probabilistic model based control scheme a reasonable candidate for application in large practical power systems, where the chances of convergence to a single model are remote. Moreover during the fault, the dynamics of the system is represented in a more realistic way by a combination of several models rather than by a single model

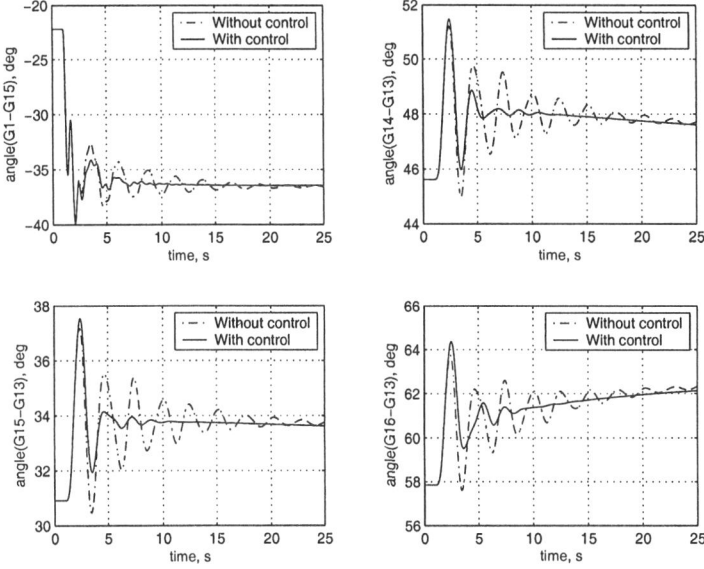

Figure 6.21. Test Case IIb: Dynamic response of the system

6.10 Summary

In this chapter, the application of a multiple-model adaptive control scheme for robust damping of inter-area oscillations in power system using a TCSC is demonstrated. The lack of robustness of the conventional controllers under varying operating conditions leads to the motivation behind adopting such an adaptive strategy. A recursive Bayesian approach is used for computing the current probability of each model being close to the post-disturbance behavior of the system and the results are used to determine the subsequent control actions. The control output of each individual controller is assigned a weight based on the computed probability of each model and the resulting control action is the probability-weighted average of the control moves of individual controllers. The algorithm is shown to work satisfactorily for the study system under two different test cases where the model corresponding to the post-disturbance behavior is either present or not present in the model bank. When the model is present, the recursive Bayesian approach is able to identify the proper model within a few iterative steps and switch to the appropriate controller accordingly. On the other hand, when the exact model is removed from the bank, the scheme performs an appropriate blending of the remaining control moves to achieve reasonably similar performance as before. This highlights the potential applicability of the MMAC scheme for large practical power systems where the dynamics are unlikely to be governed by a single model. Under such situations, the key to the success of the MMAC scheme is the rate of convergence of the probabilities, which in turn, is governed by the proper choice of convergence factor C_f and artificial cut-off β_{min}. This chapter provides an outline on the variation pattern of the computed weights for different values of C_f and β_{min} and attempts to set a tentative guideline for choosing them, depending on the situation.

References

[sim, 2002] (2002). *Using Simulink*. The Math Works Inc., USA.

[Athans et al., 1977] Athans, M., Castanon, D., Dunn, K. P., Greene, C.S., Lee, W.H., Sandell, N.S., and Willsky, A.S. (1977). The stochastic control of the F-8C aircraft using a multiple-model adaptive control MMAC method-part I : Equilibrium flight. *IEEE Transactions on Automatic Control*, AC-22(5):768–780.

[Bandyopadhaya and Prabhu, 1988] Bandyopadhaya, G. and Prabhu, S.S. (1988). A new apporach to adaptive power system stabilizers. *Electric Machines and Power Systems*, 14:111–125.

[Chaudhuri et al., 2004] Chaudhuri, B., Majumder, R., and Pal, B.C. (2004). Application of multiple-model adaptive control strategy for robust damping of interarea oscillations in power system. *IEEE Transactions on Control System Technology*, 12(5):727–736.

[He et al., 1986] He, W.G., Kaufman, H., and Roy, R. (1986). Multiple-model adaptive control procedure for blood pressure control. *IEEE Transactions on Biomedical Engineering*, BME-33:10–19.

[Kundur, 1994] Kundur, P. (1994). *Power System Stability and Control*. McGraw Hill, USA.

[Kuo and Golnaraghi, 2003] Kuo, B.C. and Golnaraghi, F. (2003). *Automatic Control Systems*. John Wiley and Sons, USA.

[Lainiotis, 1976] Lainiotis, D.G. (1976). Partitioning: A unifying framework for adaptive systems, II: Control. *Proceedings of IEEE*, 64:1182–1198.

[Malik et al., 1976] Malik, O.P., Hope, G.S., and Ramanujan, V. (1976). Real time model reference adaptive control of synchronous machine excitation. *IEEE PES Winter Meeting*, 178:297–304.

[Martin et al., 1987] Martin, J.F., Schneider, A.M., and Smith, N.T. (1987). Multiple-model adaptive control of blood pressure using sodium nitroprusside. *IEEE Transactions on Biomedical Engineering*, BME-34:603–611.

[Martins and Lima, 1990] Martins, N. and Lima, L.T.G. (1990). Determination of suitable locations for power system stabilizers and static var compensators for damping electromechanical oscillations in large power systems. *IEEE Transactions on Power Systems*, 5(4):1455–1469.

[M.Klein et al., 1991] M.Klein, Rogers, G.J., and Kundur, P. (1991). A fundamental study of inter-area oscillations in power systems. *IEEE Transactions on Power System*, 6(3):914–921.

REFERENCES

[Pahalawaththa et al., 1991] Pahalawaththa, N.C., Hope, G.S., and Malik, O.P. (1991). Multivariable self-tuning power system stabilizer simulation and implementation studies. *IEEE Transactions on Energy Conversion*, 6:310–316.

[Paserba, 1996] Paserba, J. (1996). Analysis and control of power system oscillation. *CIGRE Special Publication 38.01.07*, Technical Brochure 111.

[Rao et al., 2003] Rao, R.R., Aufderheide, B., and Bequette, B.W. (2003). Experimental studies on multiple-model predictive control for automated regulation of hemodynamic variables. *IEEE Transactions on Biomedical Engineering*, 50(3):277–288.

[Yu et al., 1992] Yu, C., Roy, R.J., Kaufman, H., and Bequette, B.W. (1992). Mutiple-model adaptive predictive control of mean arterial pressure and cardiac output. *IEEE Transactions on Biomedical Engineering*, 39(8):765–778.

Chapter 7

SIMULTANEOUS STABILIZATION

In the previous chapter, an adaptive control scheme was illustrated involving a number of controllers designed and tuned corresponding to a range of operating conditions. In this chapter, the focus is on designing a single controller which is able to guarantee the performance specification for a range of operating conditions.

The concept of robust and low-order controller design by weighted and normalized eigenvalue-distance minimization (WNEDM) is applied for improving the damping of the inter-area modes. The technique was applied for a single-input, single-output (SISO) power system model in [Pal et al., 2000, Pal, 1999] and was later extended for MISO systems in [Pal et al., 2004]. The basic idea is to place the closed-loop eigen values of the system at certain desired locations in the left-half of the complex plane. The robustness issue is addressed by considering a family of operating conditions and optimizing over the worst case scenario. To start with, the design procedure is presented for a SISO system for easy understanding. Later on it is generalized in the multi-variable framework. A case study is presented based on the power system model described in Chapter 4. The objective is to provide additional damping to three inter-area modes by a single FACTS device employing remote signals. The problem is formulated to address multi-input-single-output (MISO) control design for a group of single-input-multi-output (SIMO) system models.

7.1 Eigen-Value-Distance Minimization

In robust low-order controller design [Schmitendorf and Wilmers, 1991], the desired closed-loop pole locations are specified and a suitable controller is sought to move the open-loop poles towards the specified locations. An optimization problem is then solved with the controller parameters as the design variables. The order of the controller is assigned beforehand. For simplicity, let us begin with a SISO system model. The general feedback control set-up for such a system is shown in Fig. 7.1 where, $G(s)$ and $K(s)$ are the transfer functions of the system and the controller, respectively.

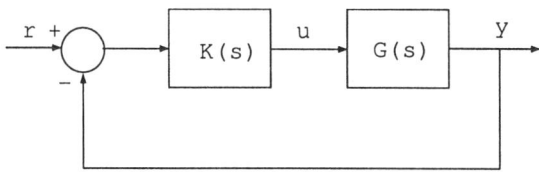

Figure 7.1. Closed-loop feedback configuration with negative feedback

In polynomial form, an nth order system $G(s)$ and a mth order controller $K(s)$ can be represented as follows:

$$G(s) = \frac{N_g(s)}{D_g(s)} = \frac{n_{gn}s^n + n_{gn-1}s^{n-1} + \ldots + n_{g0}}{d_{gn}s^n + d_{gn-1}s^{n-1} + \ldots + d_{g0}} \quad (7.1)$$

$$K(s) = \frac{N_k(s)}{D_k(s)} = \frac{n_{km}s^m + n_{km-1}s^{m-1} + \ldots + n_{k0}}{d_{km}s^m + d_{km-1}s^{m-1} + \ldots + d_{k0}} \quad (7.2)$$

The transfer functions are proper if $d_{gn} \neq 0$ and $d_{km} \neq 0$. The closed-loop transfer function is given by:

$$T(s) = \frac{G(s)K(s)}{1 + G(s)K(s)} = \frac{N_g(s)N_k(s)}{N_g(s)N_k(s) + D_g(s)D_k(s)} \quad (7.3)$$

The characteristic polynomial for (7.3) is given by:

$$\delta(s) = N_g(s)N_k(s) + D_g(s)D_k(s) \quad (7.4)$$
$$\equiv \delta_{n+m}s^{n+m} + \delta_{n+m-1}s^{n+m-1} + \ldots + \delta_1 s + \delta_0$$

where, $\delta = [\ \delta_{n+m} \ \ \delta_{n+m-1} \ \ \cdot \ \ \cdot \ \ \cdot \ \ \delta_1 \ \ \delta_0\ ]$ is the closed-loop characteristic vector. The roots of (7.4) are the closed-loop poles of the system. The coefficients of the transfer function of the controller in (7.2) can be reorganized in the form of a vector as follows:

$$\mathbf{x} = [\ n_{km} \ \ d_{km} \ \ n_{km-1} \ \ d_{km-1} \ \ \cdot \ \ \cdot \ \ \cdot \ \ n_{k1} \ \ d_{k1} \ \ n_{k0} \ \ d_{k0}\ ] \quad (7.5)$$

The vector $\mathbf{x}$ is known as the controller parameter vector. Equating the coefficients of equal power of s on both sides of (7.4) and using (7.5), the following equation is obtained [Chen, 1999].

Simultaneous Stabilization

$$Px = \delta \tag{7.6}$$

where, P is a $(n+m+1) \times (2m+2)$ matrix with the following structure:

$$P = \begin{bmatrix} n_{gn} & d_{gn} & 0 & 0 & \cdots & 0 & 0 \\ n_{gn-1} & d_{gn-1} & n_{gn} & d_{gn} & \cdots & 0 & 0 \\ \cdot & \cdot & \cdot & \cdot & \cdots & \cdot & \cdot \\ n_{g0} & d_{g0} & n_{g1} & d_{g1} & \cdots & n_{gn} & d_{gn} \\ 0 & 0 & n_{g0} & d_{g0} & \cdots & n_{gn-1} & d_{gn-1} \\ \cdot & \cdot & \cdot & \cdot & \cdots & \cdot & \cdot \\ \cdot & \cdot & \cdot & \cdot & \cdots & n_{g1} & d_{g1} \\ \cdot & \cdot & \cdot & \cdot & \cdots & n_{g0} & d_{g0} \end{bmatrix} \tag{7.7}$$

The matrix P is called the plant parameter matrix as its elements are obtained from the coefficients of the system model or plant polynomial in (7.1).

Any δ, hence an arbitrary pole-placement, can be achieved through the proper choice of x, if δ is in the column space of P. However, this is possible only when the matrix P is of full row-rank i.e. when $m \geq n-1$. Therefore, for arbitrary pole-placement, the minimum order of the controller $K(s)$ has to be just one less than that of the system itself i.e. $n-1$.

For a large order system, such as a power system, the order of the controller becomes large. It can be reduced if the aim of the arbitrary pole-placement for all the closed-loop poles is relaxed to closed-loop stability with special emphasis on a few critical poles. Incidentally, this is exactly what is required in power system-damping controller design where damping ratios of a few critical electromechanical modes are of importance as long as the others are stable.

It is extremely difficult, if not impossible, to find a reduced order $m < n-1$ controller (i.e. vector of controller parameters x), such that (7.6) is satisfied for a desired characteristic polynomial δ^*. The best that can be done is to choose x such that $|Px - \delta^*|$ is minimized. A solution to this optimization problem brings the actual characteristic polynomial close to the desired one but the closed-loop poles might not be near those desired. Therefore, it is preferable to minimize the distance between the desired and the actual eigenvalues.

A straightforward distance minimization approach suffers from certain drawbacks. For example, given that two of the desired real poles are at -15.0 and -2.0, if the optimization algorithm provides a solution of -12.5 and -0.5, the absolute error of 1.5 is the same for both the poles. But the pole at -0.5 causes a larger change in the dynamics of the system. In view of this, it is necessary to minimize the normalized eigenvalue-distance instead of minimizing absolute eigenvalue-distance. In order to attach more importance to the critical eigenvalues, the objective of eigenvalue-distance minimization must be weighted in addition to normalization. To achieve this, the modified objective function becomes:

$$F(\mathbf{x}) = \sum_{i=1}^{n+m} c_i \frac{\|\mu_i - \lambda_i(\mathbf{x})\|}{\|\mu_i\|} \tag{7.8}$$

where, μ_i and $\lambda_i(\mathbf{x})$ is the desired and actual locations of the ith closed-loop pole and c_i is the weight associated with it. The controller parameter vector $\mathbf{x}$ is optimized based on the plant parameter matrix P.

Proper selection of weights c_i is extremely important for the optimization algorithm to produce desired results. For unstable poles, real poles with a small absolute magnitude and poorly damped complex poles, the weights must be very high. For other poles, the weights can be relatively smaller. For any unstable open-loop pole, $c_i = 1000$ is a recommendable choice. For real poles with a small absolute magnitude, the c_i can be chosen to be between 10 to 100. Any poorly damped complex poles can be weighted by $c_i = m_c \frac{\zeta_{min}}{\zeta}$; where, ζ_{min} is the minimum required damping ratio of the pole, ζ is the damping ratio as the iteration proceeds and m_c is an appropriately selected constant. The choice of suitable weights are suggested in detail in [Schmitendorf and Wilmers, 1991].

The concept of weighted and normalized eigen-value-distance-minimization approach for a SISO system is extendable to any single-input, multi-output (SIMO) system. Let us consider a 1-input, 3-output system and a 3-input, 1-output controller as shown in Fig. 7.2. Note that the feedback sense in this case is chosen to be positive and the controller is present in the feedback path. This is used to illustrate that the formulation is general and can be extended to different possible configurations of feedback control.

The system model $G(s)$ and the controller $K(s)$ are given by:

$$G(s) = \begin{bmatrix} G_1(s) \\ G_2(s) \\ G_3(s) \end{bmatrix} = \frac{1}{D_g(s)} \begin{bmatrix} N_{g,1}(s) \\ N_{g,2}(s) \\ N_{g,3}(s) \end{bmatrix} \tag{7.9}$$

$$K(s) = \begin{bmatrix} K_1(s) \\ K_2(s) \\ K_3(s) \end{bmatrix}^T = \frac{1}{D_k(s)} \begin{bmatrix} N_{k,1}(s) \\ N_{k,2}(s) \\ N_{k,3}(s) \end{bmatrix}^T \tag{7.10}$$

where $N_{g,i}(s), D_g(s), N_{k,i}(s)$ and $D_k(s)$ are given by

Simultaneous Stabilization

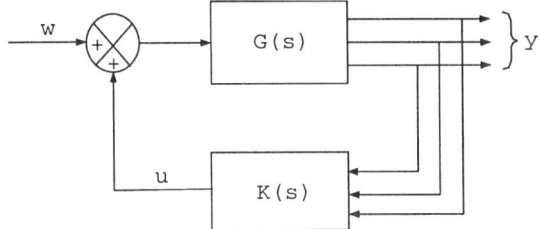

Figure 7.2. Closed-loop feedback configuration for a 1-input, 3-output system with positive feedback

$$N_{g,i}(s) = n_{gn,i}s^n + n_{gn-1,i}s^{n-1} + .. + n_{g1,i}s^1 + n_{g0,i}s^0 \qquad (7.11)$$
$$D_g(s) = d_{gn}s^n + d_{gn-1}s^{n-1} + .. + d_{g1}s^1 + d_{g0}s^0 \qquad (7.12)$$
$$N_{k,i}(s) = n_{km,i}s^m + n_{km-1,i}s^{m-1} + .. + n_{k1,i}s^1 + n_{k0,i}s^0 \qquad (7.13)$$
$$D_k(s) = d_{km}s^m + d_{km-1}s^{m-1} + .. + d_k1s^1 + dk_0s^0 \qquad (7.14)$$

for i = 1 to 3. The closed-loop transfer vector is expressed as

$$G_{cl}(s) = \left[\frac{D_k(s) \begin{bmatrix} N_{g,1}(s) \\ N_{g,2}(s) \\ N_{g,3}(s) \end{bmatrix}}{\delta(s)} \right] \qquad (7.15)$$

where,

$$\delta(s) = D_k(s)D_g(s) - \sum_{i=1}^{3} N_{g,i}(s)N_{k,i}(s) \qquad (7.16)$$

is the closed-loop characteristic polynomial. The elements of δ are the coefficients of the characteristic polynomial. A vector $\mathbf{x}$ comprising the negatives of the coefficients of numerator and denominator polynomials of the controller is defined as follows:

$$\mathbf{x} = \begin{bmatrix} -n_{km,1} & -n_{km,2} & -n_{km,3} & d_{km} & \cdot & \cdot & -n_{k0,1} & -n_{k0,2} & -n_{k0,3} & b_{k0} \end{bmatrix}^T \qquad (7.17)$$

The negative sign before each entry of the coefficients in the numerator of the controller in (7.17) takes care of positive feedback without disturbing the plant parameter matrix P which in this case is given by a $(n+m+1) \times 4(m+1)$ matrix. In general, for the 1-input, r-output case, the dimension of P is $(n+m+1) \times (r+1)(m+1)$. Having formulated the plant parameter matrix, the rest of the algorithm is similar to that outlined for the SISO case.

7.2 Robust pole-placement

In the previous section, we have seen that the controller parameter vector x is optimized based on the plant parameter matrix P. With change in operating conditions, the plant parameter matrix P varies. In order to maintain acceptable performance for several operating conditions, a robustness property has to be built into the controller. This is achieved by extending the technique, described in the previous section, to incorporate other plant parameter matrices P corresponding to a number of (say q) probable operating conditions.

For a specific δ and x, q weighted and normalized eigenvalue-distances $F_j(\mathbf{x})$ are evaluated corresponding to q different P matrices. The ultimate aim of the optimization procedure is to find an x such that the maximum of all of the objective functions is minimized. The weights c_i are chosen such that the poorly damped modes are penalized heavily for the worst case $F_j(\mathbf{x})$.

This is an unconstrained optimization problem as there is no bound on x which might produce an unstable controller to provide the desired closed-loop damping. To overcome this possibility a constraint is imposed to ensure a stable controller. The controller design task, therefore, reduces to a constrained non-linear optimization problem given by:

$$\min_{\mathbf{x}} \max_{j} F_j(\mathbf{x})$$
$$sub : real\{roots(D_k(s))\} < 0 \quad (7.18)$$

The initial value of x can be chosen at random but a least-square solution of (7.6) provides a better initial guess for x. This constrained *min-max* problem is best solved by a Sequential Quadratic Programming (SQP) [Brayton et al., 1979, Fletcher, 1980] technique using the the *Optimization Toolbox* [Coleman et al., 2001] available in *Matlab* [mat, 1998]. The SQP has three stages:

1 Updating of the Hessian matrix of the Lagrangian function

2 Quadratic programming problem solution

3 Line search and merit function calculation

In each iteration, a positive definite quasi-Newton approximation of the Hessian of the Lagrangian function is calculated using the Broyden-Fletcher-Goldfarb-Shano (BFGS) [Broyden, 1970, Fletcher, 1970, Goldfarb, 1970, Shanno, 1970] method. In each iteration, a local quadratic approximation is used to compute an optimal search direction with the updated Hessian matrix. The optimal search direction is found as a solution to the local quadratic programming problem. The projection method suggested in [Gill et al., 1991] is used. Having found the optimal search direction, the optimum search length is determined by

Simultaneous Stabilization 111

another single variable optimization method that utilizes a line search strategy with an objective to minimize a merit function. Out of a number of available merit functions, the one suggested in [Dantzig, 1963] seems to work very well when the original cost functions and constraints are highly non-linear in nature. This also ensures large contributions to the penalty parameter from constraints with smaller gradients, which would be the case for active constraints at the solution point.

7.3 Case study

In this section, the prototype power system model, described in Chapter 4, is considered to illustrate the control design methodology in details. The performance and robustness of the design is also validated.

7.4 Control design

To facilitate control design and to reduce the complexity of the designed controller, the nominal system model was simplified to a 7th order equivalent as described in Chapter 4. The plant parameter matrices were formed from the simplified system models corresponding to the following operating conditions:

- All the tie-lines between NETS and NYPS in place.

- Outage of one of the tie-lines connecting buses 53 and 54

- Outage of one of the tie-lines connecting buses 60-61

- Outage of one of the tie-lines connecting buses 27-53

A converged solution, x_{opt} was considered acceptable if the damping ratios for the critical inter-area modes were more than 0.15 for all four operating conditions. Initially, we attempted the design with a second order controller but it could not produce desirable damping for all the critical modes. The controller order was increased until the desired damping to all three critical modes was achieved. The resulting controller was of 6th order.

The *fminimax* function available in the *Optimization Toolbox* [Coleman et al., 2001] in Matlab [mat, 1998] was used to find the parameters of the controller. The weights were selected as 1000 for real poles with very small decay rates and $10000 \frac{0.12}{\rho}$ for the poles with poor damping ratios where, ρ is the damping ratio calculated in every iterative step. The optimization process converged within 10 iterations. Table 7.1 lists the specified and achieved closed-loop poles for the reduced system model. The poles shown in boldface correspond to the equivalent critical inter-area modes of the reduced order system.

Table 7.1. Specified and achievable pole locations for the reduced closed-loop system

Target Pole Locations	Achieved Pole Locations
-70.0	-59640.0
-50.0	-1264.3
-20.0	-328.62
-2.0	-30.815
$-1.44 \pm j\, 3.1285$	$-2.48 \pm j\, 3.3076$
$-0.42919 \pm j\, 6.7477$	$-0.42525 \pm j\, 6.7449$
$-0.16726 \pm j\, 0.18424$	$-0.19371 \pm j\, 0.20403$
$-0.55963 \pm j2.0849$	**$-0.51534 \pm j2.1137$**
$-0.31778 \pm j3.0071$	**$-0.2715 \pm j2.9742$**
$-0.39049 \pm j3.8164$	**$-0.36348 \pm j3.8327$**

7.5 Simulation results

To evaluate the performance and robustness of the designed controller, simulations were carried out in *Simulink* [sim, 2002] for 25 s employing the *trapezoidal* integration method with a variable step size. A three-phase solid fault for 80 ms (5 cycles) was simulated followed by auto-reclosing of the breaker or outage of a tie-line. The dynamic response of the system following the disturbance is shown in Fig. 7.3.

The figures exhibit the relative angular separation between the generators located in separate geographical regions. It can be seen that inter-area oscillation settles within the specified time of 15 s for a range of post-fault operating conditions and thus abides by the robustness requirement as well. A hard limit of 0.1 to 0.8 was imposed on the percentage compensation variation of the TCSC which is depicted in Fig. 7.4.

7.6 Summary

In this chapter, the concept of simultaneous stabilization in the multi-variable framework has been presented. An optimization problem was solved to determine the controller parameters that would guarantee closed-loop poles in certain target locations with preferential treatment to those poles corresponding to the inter-area modes. System models under different operating conditions were incorporated into the design formulation to achieve performance robustness. A min-max approach was adopted to optimize the worst case scenario. The design methodology has been applied for multiple swing mode damping through a single FACTS device.

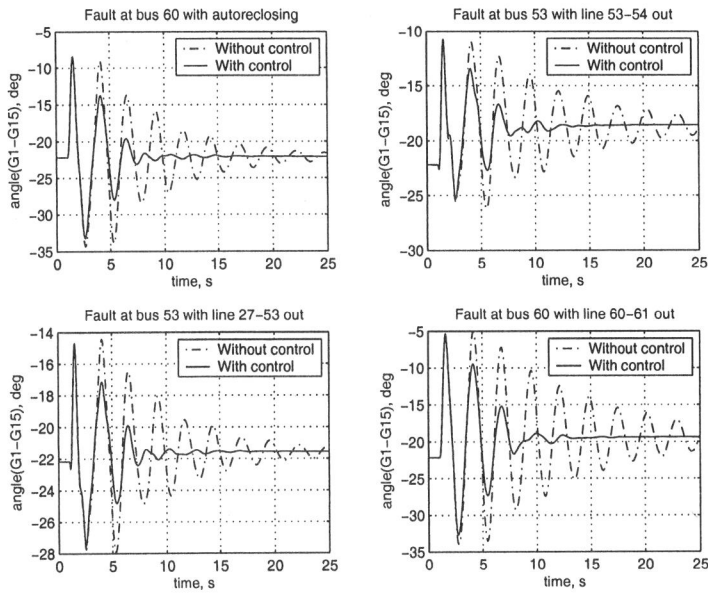

Figure 7.3. Dynamic response of the system

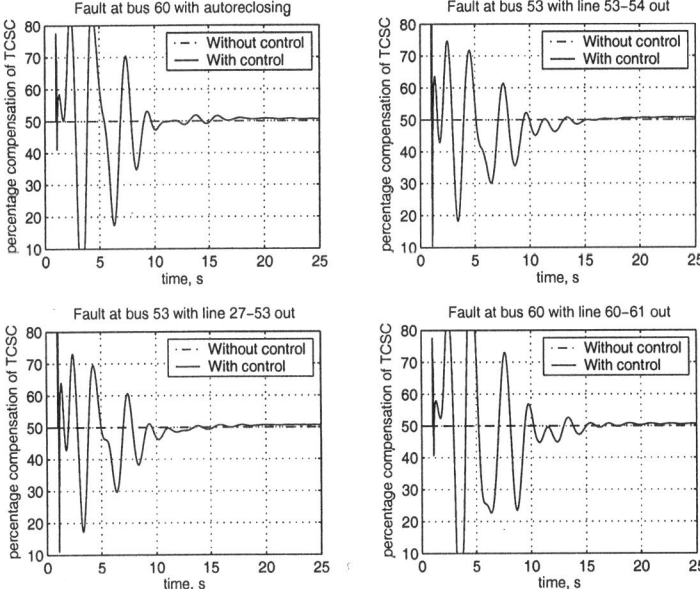

Figure 7.4. Dynamic response of the system

References

[mat, 1998] (1998). *Matlab Users Guide*. The Math Works Inc., USA.

[sim, 2002] (2002). *Using Simulink*. The Math Works Inc., USA.

[Brayton et al., 1979] Brayton, R.K., Director, S.W., Hachtel, G.D., and Vidigal, L. (1979). A new algorithm for statistical circuit design based on quasi-newton methods and function splitting. *IEEE Transactions on Circuits and Systems*, 26(9):784–794.

[Broyden, 1970] Broyden, C.G. (1970). The convergence of a class of double-rank minimization algorithms. *Journal of Institute of Maths Applications*, 6:76–90.

[Chen, 1999] Chen, Chi-Tsong (1999). *Linear System Theory and Design*. Oxford University Press, New York.

[Coleman et al., 2001] Coleman, T., Branch, M.A., and Grace, A. (2001). *Optimization Toolbox For Use With MATLAB*. The Math Works Inc., USA.

[Dantzig, 1963] Dantzig, G. (1963). *Linear Programming and Extensions*. Princeton University Press.

[Fletcher, 1970] Fletcher, R. (1970). A new approach to variable metric algorithms. *Computer Journal*, 13:317–322.

[Fletcher, 1980] Fletcher, R. (1980). *Practical Methods of Optimization, vol. 2*. John Wiley and Sons.

[Gill et al., 1991] Gill, P.E., Murray, W., and Wright, M.H. (1991). *Numerical linear Algebra and Optimization, vol. 1*. Addison Wesley.

[Goldfarb, 1970] Goldfarb, D. (1970). A family of variable metric updates derived by variational means. *Mathematics of Computing*, 24:23–26.

[Pal, 1999] Pal, B.C. (1999). *Robust Damping Control of Inter-area Oscillations in Power System with Super-conducting Magnetic Energy Storage Devices*. PhD thesis, Imperial College of Science Technology and Medicine, Department of Electrical and Electronic Engineering.

[Pal et al., 2004] Pal, B.C., Chaudhuri, B., Zolotas, A.C., and Jaimoukha, I.M. (2004). Simultaneous stabilization approach for power system damping control design through tcpar employing global signals. *IEE Proceedings: Generation, Transmission and Distribution*, 151(1):43–50.

[Pal et al., 2000] Pal, B.C., Coonick, A.H., and Macdonald, D.C. (2000). Robust damping controller design in power systems with superconducting magnetic energy storage devices. *IEEE Transactions on Power Systems*, 15(1):320–325.

[Schmitendorf and Wilmers, 1991] Schmitendorf, W.E. and Wilmers, C. (1991). Simultaneous stabilization via low order controllers. *Control and Dynamic Systems*, 35(1):165–184.

[Shanno, 1970] Shanno, D.F. (1970). Conditioning of quasi-newton methods for function minimization. *Mathematics of Computing*, 24:647–656.

Chapter 8

MIXED-SENSITIVITY APPROACH USING LINEAR MATRIX INEQUALITIES

8.1 Introduction

Inter-area oscillations in power systems are triggered by disturbances such as variation in load demand, action of voltage regulator due to a short circuit, etc.. The primary function of the damping controllers is to minimize the impact of these disturbances on the system. At the same time, it has to be ensured that the control effort is optimized considering the practical rating of the actuator devices (excitation systems, FACTS devices). In $\mathcal{H}_\infty$ control terminology, this is equivalent to designing a controller that minimizes the infinity norm of a chosen mix of closed-loop quantities as elaborated in this chapter.

Application of $\mathcal{H}_\infty$ techniques for power system damping control design has been reported in the literature to guarantee stable and robust operation of the system [Klein et al., 1995, Zhao and Jiang, 1995, Kamwa et al., 2000, Taranto and Chow, 1995]. An interesting comparison between various techniques is made in [Boukarim et al., 2000]. There are two approaches for solving a standard $\mathcal{H}_\infty$ optimization problem. One is the analytical approach wherein a positive semi-definite solution to the Ricatti equation [Skogestad and Postlethwaite, 2001] is sought. Another approach is to numerically optimize certain performance index such that Riccati inequality is satisfied. Although the Riccati inequality is non-linear, there are linearization techniques to convert it into linear matrix inequalities (LMIs) [Zhou et al., 1995, Skogestad and Postlethwaite, 2001] which are easier to handle computationally.

The analytical approach is relatively straightforward as it involves a non-iterative solution. However, an analytical solution to the $\mathcal{H}_\infty$ control design problem based on the Riccati equation approach generally produces a controller that suffers from pole-zero cancellations between the plant and the controller [Sefton and Glover, 1990]. Furthermore, additional design specifications such as the closed-loop damping ratio cannot be captured in a straight forward manner in a Riccati based design [Pal et al., 2000]. The numerical approach to solution using the linear matrix inequality (LMI) approach has a distinct advantage as design specifications can be addressed as additional constraints. To ensure a minimum damping ratio, the poles of the closed-loop system can be placed within a certain region of the complex plane which is known as pole-placement. For power system damping control applications, this is extremely important as, in addition to ensuring performance robustness, a minimum settling time is also mandatory. Moreover, the controllers obtained through a numerical approach do not suffer from the problem of pole-zero cancellation in general [Gahinet and Apkarian, 1994].

Application of the $\mathcal{H}_\infty$ approach using LMIs has been reported in [Rao and Sen, 2000, Taranto et al., 1998] for design of power system stablizers (PSS). A mixed-sensitivity approach with LMI based solution was applied for damping control design through super-conducting magnetic energy storage (SMES) devices [Pal et al., 2000, Pal et al., 1999, Pal et al., 2001]. Recently, this approach has been extended for damping control design through different FACTS devices [Chaudhuri et al., 2003, Chaudhuri and Pal, 2004].

This chapter elaborates on the basic concept of mixed-sensitivity design formulation with the problem translated into a generalized $\mathcal{H}_\infty$ problem [Zhou et al., 1995, Skogestad and Postlethwaite, 2001]. The solution is sought numerically using LMIs with additional pole-placement constraints. The entire control design methodology is illustrated by a couple of case studies on a prototype power system. The performance and robustness of the design is also validated using frequency domain analysis and simulations.

8.2 $\mathcal{H}_\infty$ mixed-sensitivity formulation

The standard mixed-sensitivity formulation for output disturbance rejection and control effort optimization is shown in Fig. 8.1, where, $G(s)$ is the open-loop system model, $K(s)$ is the controller to be designed.

The sensitivity $S = (I - GK)^{-1}$ represents the transfer function between the disturbance input $d(s)$ and the measured output $y(s)$. For minimizing the impact of any disturbance on the measured output i.e. for disturbance rejection, it is required to minimize $\|S\|_\infty$. It is required to minimize $\mathcal{H}_\infty$ norm of the transfer function between the disturbance input $d(s)$ and the control input $u(s)$ To optimize the control effort within a limited bandwidth. This is equivalent to minimizing $\|KS\|_\infty$. Thus, the minimization problem can be summarized as follows:

Mixed-sensitivity Approach Using LMI

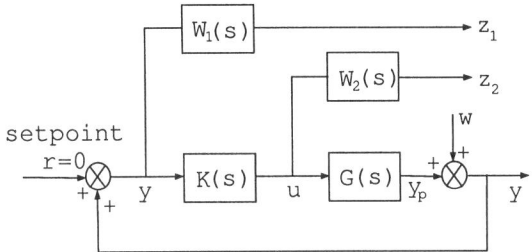

Figure 8.1. Mixed-sensitivity formulation

$$\min_{K \in \mathbb{S}} \left\| \begin{bmatrix} S \\ KS \end{bmatrix} \right\|_\infty \quad (8.1)$$

where $\mathbb{S}$ is the set of all internally stabilizing controllers K.

However, it is not possible to simultaneously minimize both S and KS over the whole frequency spectrum. This is not required in practice either. The disturbance rejection is usually required at low frequencies, thus S can be minimized over the low frequency range whereas, KS can be minimized at higher frequencies where limited control action is required. Appropriate weighting filters $W_1(s)$ and $W_2(s)$ are used to emphasize the minimization of each individual transfer function at the different frequency ranges of interest. The minimization problem is formulated such that S is less than $\frac{1}{W_1(s)}$ and KS is less than $\frac{1}{W_2(s)}$. The standard practice, therefore, is to select $W_1(s)$ as an appropriate low pass filter for output disturbance rejection and $W_2(s)$ as a high-pass filter to reduce the control effort over the high frequency range.

With the introduction of the weights, the problem can be restated as follows: Find a stabilizing controller, K, such that

$$\min_{K \in \mathbb{S}} \left\| \begin{bmatrix} W_1 S \\ W_2 KS \end{bmatrix} \right\|_\infty < 1 \quad (8.2)$$

8.3 Generalized $\mathcal{H}_\infty$ problem with pole-placement

The mixed-sensitivity problem, described in the previous section, can be solved by converting it into a generalized $\mathcal{H}_\infty$ problem. The first step is to set up a generalized regulator P corresponding to the mixed-sensitivity formulation. For simplicity, it is assumed that the weights W_1 and W_2 are not present. These can be taken care of later. Without the weights, the mixed-sensitivity formulation in Fig. 8.1 can be redrawn in terms of the A, B, C matrices of the system as shown in Fig. 8.2. Without any loss of generality, it can be assumed that $D = 0$.

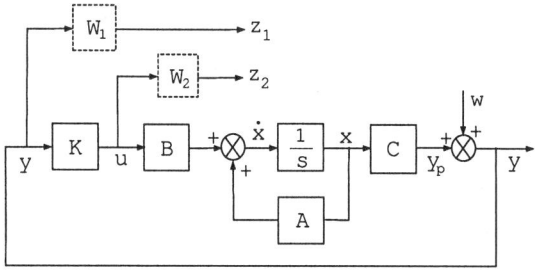

Figure 8.2. Generalized regulator set-up for mixed-sensitivity formulation

From Fig. 8.2, it can be readily seen that

$$\dot{x} = Ax + Bu \tag{8.3}$$
$$z_1 = Cx + w \tag{8.4}$$
$$z_2 = u \tag{8.5}$$
$$y = Cx + w \tag{8.6}$$

Therefore, the state-space representation of a generalized regulator P is given by:

$$\begin{bmatrix} \dot{x} \\ \hline z_1 \\ z_2 \\ y \end{bmatrix} = \begin{bmatrix} A & \| & 0 & | & B \\ \hline C & \| & I & | & 0 \\ 0 & \| & 0 & | & I \\ C & \| & I & | & 0 \end{bmatrix} \begin{bmatrix} x \\ \hline w \\ u \end{bmatrix} \tag{8.7}$$

where x: state variable vector of the power system (e.g. machine angle, machine speed etc),
w: disturbance input (e.g. a step change in excitation system reference),
u: control input (e.g. output of PSS or FACTS controllers),
y: measured output (e.g. power flow, line current, bus voltage etc),
z: regulated output.

To include the effect of the weighting filters in the generalized regulator, the state-space representations of W_1 and W_2 are placed in a diagonal form using the *sdiag* function available in Matlab [mat, 1998] and the result is multiplied with P (without the weights) using the *smult* function also available in Matlab [mat, 1998].

Mixed-sensitivity Approach Using LMI

Having formulated the generalized regulator the next task is to find an LTI control law $u = K(s)y$ for some $\mathcal{H}_\infty$ performance index $\gamma > 0$, such that: $\|T_{wz}\|_\infty < \gamma$, where, $T_{wz}(s)$ denotes the closed-loop transfer function from w to z. If the state-space representation of the LTI controller is given by:

$$\dot{x}_k = A_k x_k + B_k y \qquad (8.8)$$
$$u = C_k x_k + D_k y$$

then the closed-loop transfer function $T_{wz}(s)$ from w to z is given by $T_{wz}(s) = D_{cl} + C_{cl}(sI - A_{cl})^{-1} B_{cl}$ where,

$$A_{cl} = \begin{bmatrix} A + B_2 D_k C_2 & B_2 C_k \\ B_k C_2 & A_k \end{bmatrix} \qquad (8.9)$$

$$B_{cl} = \begin{bmatrix} B_1 + B_2 D_k D_{21} \\ B_k D_{21} \end{bmatrix} \qquad (8.10)$$

$$C_{cl} = \begin{bmatrix} C_1 + D_{12} D_k C_2 & D_{12} C_k \end{bmatrix} \qquad (8.11)$$

$$D_{cl} = D_{11} + D_{12} D_k D_{21} \qquad (8.12)$$

In addition to guaranteeing robustness by achieving $\|T_{wz}\|_\infty < \gamma$, another design requirement in power systems is to ensure that the oscillations settle within 10-15 s [Paserba, 1996]. This can be achieved if the closed-loop poles corresponding to the critical modes have a minimum damping ratio i.e. they are placed within a certain region in the left half of the complex plane. In consideration of this, the above problem statement can be modified to include the pole-placement constraint: *Find an LTI control law $u = K(s)y$ such that:*

- $\|T_{wz}\|_\infty < \gamma$

- Poles of the closed-loop system lie in $\mathcal{D}$

where, $\mathcal{D}$ defines a region in the complex plane having certain geometric shapes like disks, conic sectors, vertical/horizontal strips, etc. or intersections of these. A 'conic sector', with inner angle θ and apex at the origin is an appropriate region for power system applications as it ensures a minimum damping ratio $\zeta_{min} = \cos^{-1}\frac{\theta}{2}$ for the closed-loop poles.

8.4 Matrix inequality formulation

The bounded real lemma [Gahinet and Apkarian, 1994] and the Schur's formula for the determinant of a partitioned matrix [Skogestad and Postlethwaite,

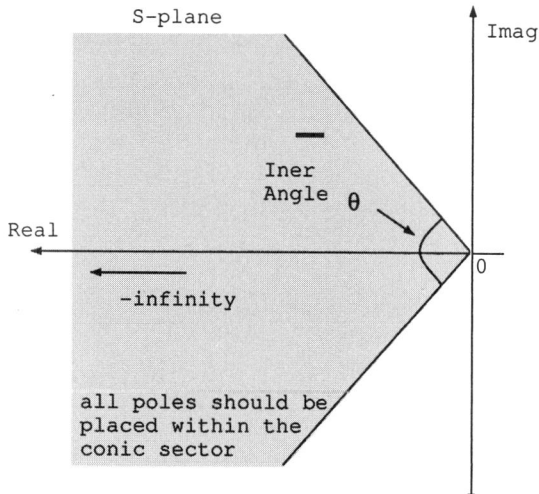

2001], enable one to conclude that the $\mathcal{H}_\infty$ constraint $\|T_{wz}\|_\infty < \gamma$ is equivalent to the existence of a solution $X_\infty = X_\infty^T > 0$ to the following matrix inequality:

$$\begin{pmatrix} X_\infty A_{cl} + A_{cl}^T X_\infty & B_{cl} & X_\infty C_{cl}^T \\ B_{cl}^T & -\gamma I & D_{cl}^T \\ C_{cl} X_\infty & D_{cl} & -\gamma I \end{pmatrix} < 0 \qquad (8.13)$$

A 'conic sector', with inner angle θ and apex at the origin is chosen as the region $\mathcal{D}$ for imposing the pole-placement constraints. The closed-loop system matrix A_{cl} has all its poles inside the conical sector $\mathcal{D}$ if and only if there exists $X_\mathcal{D} = X_\mathcal{D}^T > 0$, such that the following matrix inequality is satisfied [Scherer et al., 1996].

$$\begin{pmatrix} \sin\theta \left(A_{cl} X_\mathcal{D} + X_\mathcal{D} A_{cl}^T \right) & \cos\theta \left(A_{cl} X_\mathcal{D} - X_\mathcal{D} A_{cl}^T \right) \\ \cos\theta \left(X_\mathcal{D} A_{cl}^T - A_{cl} X_\mathcal{D} \right) & \sin\theta \left(X_\mathcal{D} A_{cl}^T + A_{cl} X_\mathcal{D} \right) \end{pmatrix} < 0 \qquad (8.14)$$

The design specifications are feasible if and only if (8.13) and (8.14) hold for some positive semi-definite matrices X_∞ and $X_\mathcal{D}$ and some controller $K = \left[\begin{array}{c|c} A_k & B_k \\ \hline C_k & D_k \end{array} \right]$. However, the problem is not jointly convex in X_∞ and $X_\mathcal{D}$ unless it is solved for the same matrix X. In view of this, the sub-optimal

Mixed-sensitivity Approach Using LMI

$\mathcal{H}_\infty$ problem with pole-placement can be stated as follows:

Find $X > 0$ and a controller K, such that (8.13) and (8.14) are satisfied with $X = X_\infty = X_\mathcal{D}$ [Chilali and Gahinet, 1997, Scherer et al., 1996].

The inequalities (8.13) and (8.14) contain $A_{cl}X$ and $C_{cl}X$. A_{cl} and C_{cl} are functions of the controller parameters A_k, B_k, C_k and D_k and the controller parameters themselves are functions of X making the products $A_{cl}X$, $C_{cl}X$ non-linear in X. A change of controller variables is necessary to convert the problem into a linear one, as described in the next section.

8.5 Linearization of the matrix inequalities

The controller variables are implicitly defined in terms of the (unknown) matrix X. Let X and X^{-1} be partitioned as:

$$X = \begin{pmatrix} R & M \\ M^T & U \end{pmatrix}, \quad X^{-1} = \begin{pmatrix} S & N \\ N^T & V \end{pmatrix} \quad (8.15)$$

For $\Pi_1 = \begin{pmatrix} R & I \\ M^T & 0 \end{pmatrix}$ and $\Pi_2 = \begin{pmatrix} I & S \\ 0 & N^T \end{pmatrix}$, X satisfies the identity $X\Pi_2 = \Pi_1$. The new controller variables are defined as:

$$\hat{A} = NA_kM^T + NB_kC_2R + SB_2C_kM^T + S(A + B_2D_kC_2)R \quad (8.16)$$
$$\hat{B} = NB_k + SB_2D_k \quad (8.17)$$
$$\hat{C} = C_kM^T + D_kC_2R \quad (8.18)$$
$$\hat{D} = D_k \quad (8.19)$$

The identity $XX^{-1} = I$ together with (8.15) gives

$$MN^T = I - RS \quad (8.20)$$

If M and N have full row rank, then the controller matrices A_k, B_k, C_k and D_k can always be computed from $\hat{A}, \hat{B}, \hat{C}, \hat{D}, R, S, M$ and N. Moreover, the controller matrices can be determined uniquely if the controller order is chosen to be equal to that of the generalized regulator [Scherer et al., 1996].

Pre- and post-multiplying the inequality $X > 0$ by Π_2^T and Π_2, respectively and carrying out appropriate change of variables according to (8.16), (8.17), (8.18) and (8.19), the following linear matrix inequality (LMI) is obtained.

$$\begin{pmatrix} R & I \\ I & S \end{pmatrix} > 0 \quad (8.21)$$

Similarly, pre- and post-multiplying the inequality (8.13) by $diag(\prod_2^T, I, I)$ and $diag(\prod_2, I, I)$, respectively; and carrying out appropriate change of variables according to (8.16), (8.17), (8.18) and (8.19), the following LMI is obtained.

$$\begin{bmatrix} \Psi_{11} & \Psi_{21}^T \\ \Psi_{21} & \Psi_{22} \end{bmatrix} < 0 \quad (8.22)$$

where,

$$\Psi_{11} = \begin{bmatrix} AR + RA^T + B_2\hat{C} + \hat{C}^T B_2^T & B_1 + B_2\hat{D}D_{21} \\ \left(B_1 + B_2\hat{D}D_{21}\right)^T & -\gamma I \end{bmatrix} \quad (8.23)$$

$$\Psi_{21} = \begin{bmatrix} \hat{A} + \left(A + B_2\hat{D}C_2\right)^T & SB_1 + \hat{B}D_{21} \\ C_1 R + D_{12}\hat{C} & D_{11} + D_{12}\hat{D}D_{21} \end{bmatrix} \quad (8.24)$$

$$\Psi_{22} = \begin{bmatrix} A^T S + SA + \hat{B}C_2 + C_2^T \hat{B}^T & \left(C_1 + D_{12}\hat{D}C_2\right)^T \\ C_1 + D_{12}\hat{D}C_2 & -\gamma I \end{bmatrix} \quad (8.25)$$

Proceeding in a similar fashion by pre- and post-multiplying the inequality (8.14) by $\prod_2^T$ and $\prod_2$, respectively, and carrying out the change of variables according to (8.16), (8.17), (8.18) and (8.19), the following LMI is obtained. Interested readers are recommended to consult [Chilali and Gahinet, 1997, Scherer et al., 1996] for details of the derivation.

$$\begin{pmatrix} \sin\theta \left(\Phi + \Phi^T\right) & \cos\theta \left(\Phi - \Phi^T\right) \\ \cos\theta \left(\Phi^T - \Phi\right) & \sin\theta \left(\Phi^T + \Phi\right) \end{pmatrix} < 0 \quad (8.26)$$

where,

$$\Phi = \begin{pmatrix} AR + B_2\hat{C} & A + B_2\hat{D}D_{21} \\ \hat{A} & SA + \hat{B}C_2 \end{pmatrix} \quad (8.27)$$

The system of LMIs in (8.21), (8.22) and (8.26) are solved for $R, S, \hat{A}, \hat{B}, \hat{C}$ and $\hat{D}$. A full-rank factorization $MN^T = I - RS$ of the matrix $I - RS$ is computed via singular value decomposition (SVD) approach such that M and N are square and invertible. With known values $R, S, \hat{A}, \hat{B}, \hat{C}, \hat{D}, M$ and N the system of linear equations (8.16), (8.17), (8.18) and (8.19) is solved for D_k, B_k, C_k and A_k in that order. The controller is obtained as $K(s) = D_k + C_k(sI - A_k)^{-1}B_k$ and the resultant controller places the closed-loop poles in $\mathcal{D}$ and satisfies $\|T_{wz}\|_\infty < \gamma$.

8.6 Case study

In this section, the prototype power system model, described in Chapter 4, is considered to illustrate the control design methodology in detail. The performance and robustness of the design is also validated.

8.6.1 Weight selection

As mentioned earlier, the standard practice in $\mathcal{H}_\infty$ mixed-sensitivity design is to choose the weight $W_1(s)$ as an appropriate low pass filter for output disturbance rejection and $W_2(s)$ as a high-pass filter to reduce the control effort in the high frequency range. In view of that, the weights are initially chosen as follows:

$$W_1(s) = \frac{30}{s+30}, \quad W_2(s) = \frac{10s}{s+100} \tag{8.28}$$

The frequency responses of these weighting functions are shown in Fig. 8.4. It can be seen that the two weights intersect at around 10 rad/s, (note that the critical modes to be controlled are below the frequency of 10 rad/s). Thus, the minimization of the sensitivity is emphasized up to this frequency and the constraint on the control effort is imposed soon after.

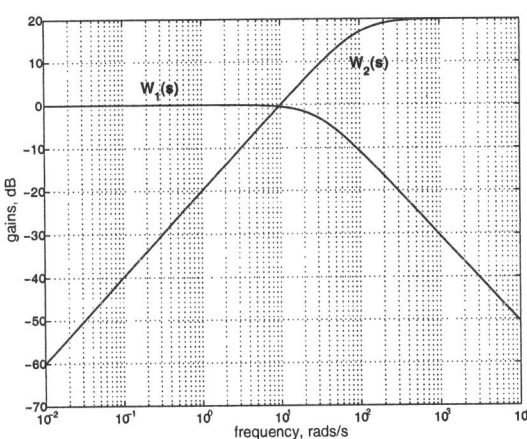

Figure 8.4. Frequency response of the weighting filters

8.6.2 Control design

To facilitate control design and to reduce the complexity of the designed controller, the nominal system model was reduced to a 7th order equivalent as described in Chapter 4. The generalized regulator problem was formulated according to (8.7) using the simplified system model and the weights in (8.28). The control design problem is to minimize γ such that (8.21), (8.22) and (8.26) are satisfied.

A series of functions which are available with the *LMI toolbox* [Gahinet et al., 1995] in *Matlab* [mat, 1998] is used to formulate and solve the optimization problem. The first step is to define the solution variables (also called the LMI variables). The variables R, S, $\hat{A}$, $\hat{B}$ and $\hat{C}$ are defined using the *lmivar* function. The size of these variables and their structure is specified through this function. Having defined the solution variables, the next step is to set up the LMIs in (8.21), (8.22) and (8.26) in terms of these variables. Each of the terms of an LMI and their respective positions are specified using the *lmiterm* function. In this design, a 'conic sector' of inner angle $2cos^{-1}0.15$ with apex at the origin was chosen as the pole-placement region to ensure a minimum damping ratio of 0.15 for the closed-loop system. To achieve this, the value of θ in (8.26) was set to $cos^{-1}0.15$.

The three sets of LMIs are combined in a system of LMI using the *getlmis* function. Once the LMIs are set up, γ is minimized using the *mincx* function such that the set of LMIs are satisfied. The optimum value of the solution variables R, S, $\hat{A}$, $\hat{B}$ and $\hat{C}$ are retrieved from the output of the *mincx* function by using the *dec2mat* function. From R and S, M and N are determined through singular value decomposition of $1 - RS$. Knowing R, S, M, N, $\hat{A}$, $\hat{B}$ and $\hat{C}$, the controller parameters A_k, B_k, C_k and D_k are determined from (8.16), (8.17), (8.18) and (8.19). The matlab code for designing a controller using the above mentioned procedure is given in the Appendix D at the end of the book.

The order of the controller obtained from this design routine is equal to the reduced system order plus the order of the weights. As there are three weights associated with the three measured outputs and one with the control input, the size of the designed controller is 14 (9+3+1). The designed controller was simplified further to a 10th order equivalent without affecting the frequency response as shown in Fig. 8.5.

The frequency response of the sensitivity S and the control times sensitivity KS is plotted in Figs. 8.6 and 8.7. As discussed before, S should be low at the lower frequencies to achieve disturbance rejection but relatively higher values can be tolerated at higher frequencies. This is achieved in the designed controller as seen from Fig. 8.6. On the other hand to ensure satisfactory performance KS should be low at high frequencies to reduce the control effort

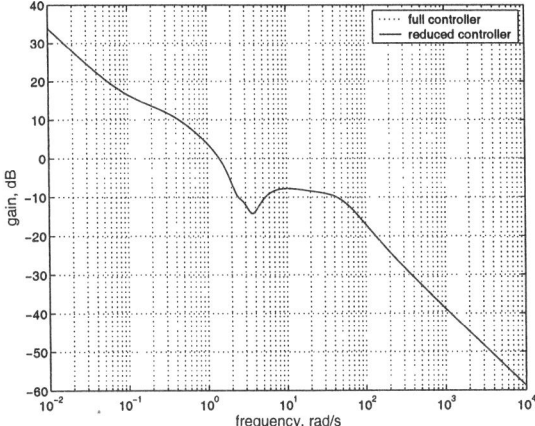

Figure 8.5. Frequency response of the full and reduced controller

which is evident in the achieved design as can be seen from Fig. 8.7.

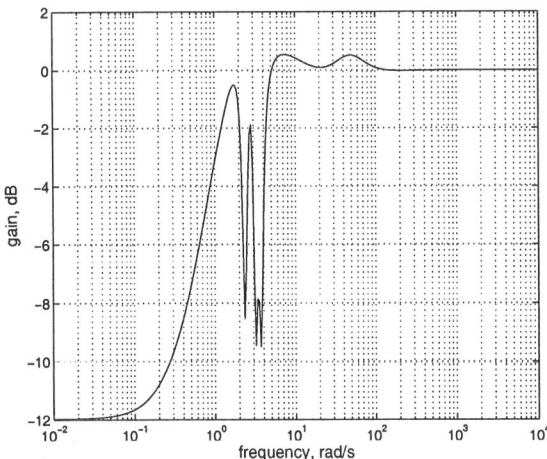

Figure 8.6. Frequency response of sensitivity (S)

The design steps can be summarized as follows:

1 Simplify the system model

2 Formulate the generalized regulator using the simplified system model and the mixed-sensitivity weights

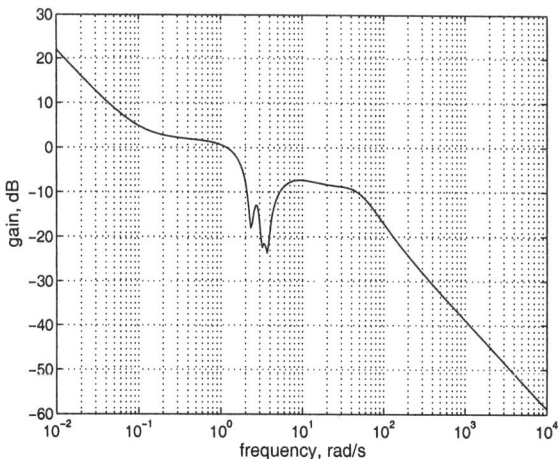

Figure 8.7. Frequency response of control times sensitivity (KS)

3 Define the LMI variables using the *lmivar* function

4 Construct the terms of the LMIs using the *lmiterm* function

5 Assemble the individual LMIs into a set of LMIs employing the *getlmis* function

6 Solve the γ optimization problem with the set of LMI constraints using the *mincx* function

7 Retrieve the optimum value of the solution variables through the *dec2mat* function

8 Determine the controller using the optimum value of the solution variables

9 Simplify the designed controller

Alternatively, the design problem can be solved by suitably defining the objectives in the argument of the function *hinfmix* available with the *LMI Toolbox* [Gahinet et al., 1995] in Matlab [mat, 1998]. The pole-placement constraint can be imposed by using the *lmireg* function which is an interactive interface for specifying different LMI regions. The matlab code illustrating the use of *hinfmix* function for this design is given in the Appendix E at the end of the book.

Table 8.1. Damping ratios and frequencies of the inter-area modes

Mode no.	No control		With control	
	ζ	$f(Hz)$	ζ	$f(Hz)$
1	0.0626	0.3913	**0.2336**	0.3590
2	0.0435	0.5080	**0.1316**	0.5094
3	0.0554	0.6232	**0.1456**	0.6384
4	0.0499	0.7915	0.0550	0.7843

8.6.3 Performance evaluation

The eigen-values of the closed-loop system were computed to examine the performance of the designed controller in terms of improving the damping ratios of the inter-area modes. The results are summarized in Tables 8.1. It can be seen that the damping ratios of the three critical inter-area modes, shown in boldface, are improved in the presence of the controller.

It is to be noted that by imposing the pole-placement constraint, as described earlier, a minimum damping ratio of 0.15 could be ensured for the simplified closed-loop system (simplified open-loop system and designed controller in feedback). However, the results shown here are based on the original system and therefore the damping ratios under certain situations are less than 0.15 but they are still adequate enough to ensure that oscillations settle within 12-15 s.

The damping action of the designed controller was examined under different types of disturbances in the system. These included changes in power flow levels over key transmission corridors, change of type of loads etc. Table 8.2 displays the damping ratios of the inter-area modes for a range of power flows across NETS and NYPS interconnection.

The performance of the controller was tested with various load models including constant impedance (CI), a mixture of constant current and constant impedance (CC+CI), a mixture of constant power and constant impedance (CP+CI) and with dynamic load characteristics. The damping ratios of the inter-area modes are listed in Table 8.3 for different types of load characteristics.

From the damping ratios displayed in the tables above, it can be concluded that action of the designed controller is robust against widely varying operating conditions.

Table 8.2. Damping ratios and frequencies of the critical inter-area modes at different levels of power flow between NETS and NYPS

Power flow (MW)	Mode 1 ζ	Mode 1 $f(Hz)$	Mode 2 ζ	Mode 2 $f(Hz)$	Mode 3 ζ	Mode 3 $f(Hz)$
100	**0.2420**	0.3566	**0.1374**	0.5106	**0.1351**	0.6640
500	**0.2371**	0.3578	**0.1338**	0.5097	**0.1419**	0.6451
700	**0.2336**	0.3590	**0.1316**	0.5094	**0.1456**	0.6384
900	**0.2300**	0.3609	**0.1289**	0.5093	**0.1491**	0.6251

Table 8.3. Damping ratios and frequencies of the critical inter-area modes for different load models

Type of load	Mode 1 ζ	Mode 1 $f(Hz)$	Mode 2 ζ	Mode 2 $f(Hz)$	Mode 3 ζ	Mode 3 $f(Hz)$
CI	**0.2336**	0.3590	**0.1316**	0.5094	**0.1456**	0.6384
CC+CI	**0.2313**	0.3608	**0.1308**	0.5175	**0.1314**	0.6353
CP+CI	**0.2251**	0.3621	**0.1309**	0.5260	**0.1175**	0.6351
Dynamic	**0.2304**	0.3582	**0.1399**	0.5135	**0.1456**	0.6381

8.6.4 Simulation results

One of the most severe disturbances stimulating poorly damped inter-area oscillations is a three-phase fault in one of the key transmission circuits. For temporary faults, the circuit breaker 'auto-recloses' and normal operation is restored, otherwise, one or two lines might have to be taken out for maintenance. There might be other types of disturbances in the system such as change of load characteristics, sudden change in power flow etc. which are less severe compared to faults and are, therefore, not considered here.

To evaluate the performance and robustness of the designed controller simulations were carried out corresponding to some of the probable fault scenarios in the NETS and NYPS inter-connection. There are three inter-connections between NETS and NYPS connecting buses #60-#61, #53-#54 and #27-#53, respectively. Each of these inter-connections consists of two lines and an outage of one of these lines weakens the interconnection considerably. The following disturbances were considered for simulation with a three-phase solid fault for 80 ms (about 5 cycles) close to the following:

1 bus #60 followed by auto-reclosing of the circuit breaker

Mixed-sensitivity Approach Using LMI 129

2. bus #53 followed by outage of one of the tie-lines between buses #53-#54
3. bus #53 followed by outage of one of the tie-lines between buses #27-#53
4. bus #60 followed by outage of one of the tie-lines between buses #60-#61

The designed controller is supposed to settle the inter-area oscillations within 12-15 s (performance criteria) following the disturbances. Moreover, it should be able to achieve this following any of the above disturbances (robustness) although the design is based on a nominal operating condition (no outage).

Simulations were carried out in Matlab *Simulink* [sim, 2002] for 25 s employing the *trapezoidal integration* method with a variable step size. The disturbance was created 1 s after the start of the simulation. The dynamic response of the system following the disturbance is shown in Figs. 8.8, 8.9 and 8.10.

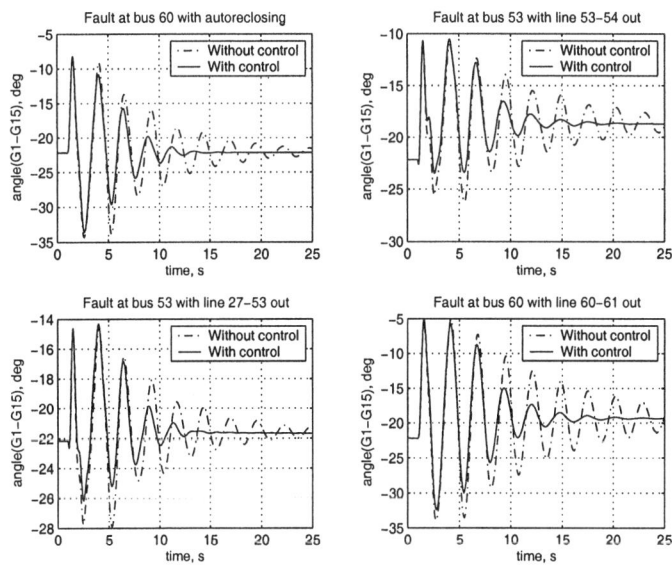

Figure 8.8. Dynamic response of the system

These figures exhibit the relative angular separation between the generators located in separate geographical regions. Inter-area oscillations are mostly manifested in these angular differences and are therefore, chosen for displaying. It can be seen that inter-area oscillations settle within the desired performance specification of 12-15 s for a range of post-fault operating conditions and thus abides by the robustness requirement as well. A hard limit of 0.1 to 0.8 was imposed on the variation of the percentage compensation of the TCSC which is depicted in Fig. 8.10.

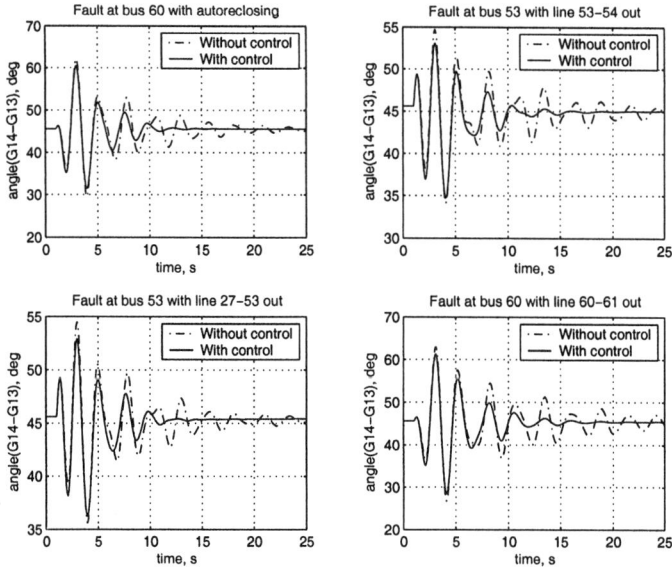

Figure 8.9. Dynamic response of the system

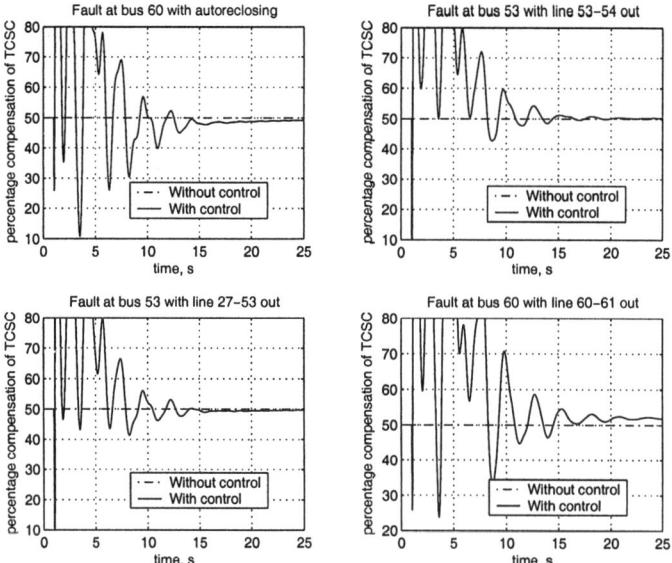

Figure 8.10. Dynamic response of the system

Mixed-sensitivity Approach Using LMI

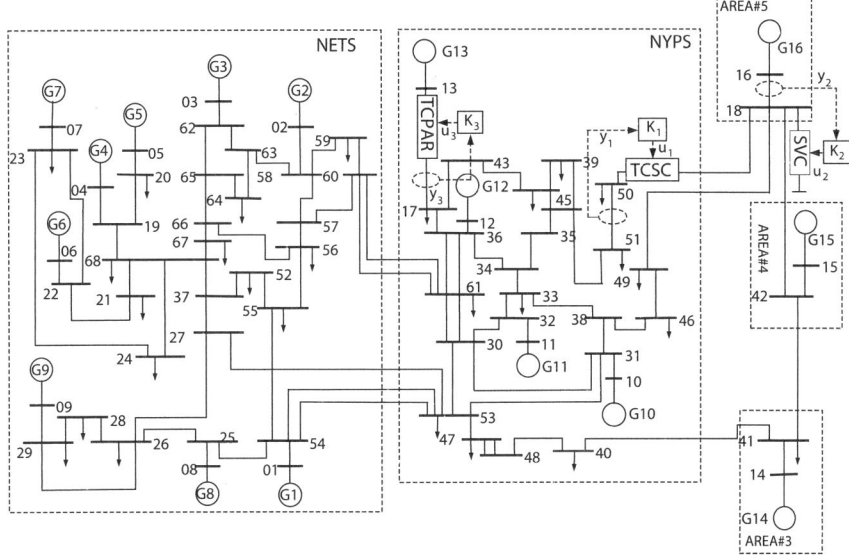

Figure 8.11. Sixteen machine five area study system with three FACTS devices

8.7 Case study on sequential design

In this section, a case study on sequential design of damping controllers for multiple FACTS devices is presented. The basic control design formulation is exactly the same as described in the previous section. However, a separate controller is designed for each of the FACTS devices sequentially. The feedback signals are chosen appropriately out of those locally available.

8.7.1 Test system

The test system used for this study is the same as described in Chapter 4. However, instead of one, three FACTS devices are considered to be installed as shown in Fig. 8.11.

The TCSC is installed in the line between buses #18 and #50 to provide a compensation (k_c) of 50%. An SVC is present at bus #18 to provide voltage support in the face of 1500 MW power transfer between area #5 and NYPS. The setting of the SVC is set to 117 MVAr to ensure nominal voltage at bus #18. A TCPAR, with a steady state phase angle (ϕ) setting of 10 degrees, is installed in the line connecting buses #13 and #17 to facilitate 3000 MW power transfer from equivalent generation G3 to the rest of the NYPS.

The aim of this exercise is to design three separate damping controllers K_1, K_2 and K_3 using locally available signals such that inter-area oscillations can be damped. The location of the FACTS devices and the corresponding damping controllers are shown in Fig. 8.11, where y_1, y_2 and y_3 are the measured feedback signals and u_1, u_2 and u_3 are the derived control signals.

8.7.2 Control design

The control design formulation described in Section 8.6.2 produces centralized controllers in multi-variable form. Here, the design of the damping controllers is done in a sequential manner using a decentralized approach. The basic idea is to design a damping controller for one device to start with. The closed-loop system using this controller is used to design the controller for the second device. Exactly the same procedure is repeated for the third device. At each stage of this sequential design, the system model is updated with the designed controller model. In such a design method, the order of the system increases as each loop is closed depending on the number of states associated with the controllers of the individual FACTS devices.

The sequential design of the controllers K_1, K_2 and K_3 for the TCSC, SVC and TCPAR has been carried out in sequence. The choice of this sequence improves the damping of modes #1, #2 and #3 in that order. Other sequences were tested and found to produce slightly different controllers but essentially the same performance could be achieved. The same set of weights given in (8.29) and (8.29) has been found to work well for the design of all three controllers.

$$W_1(s) = 0.8475 \frac{99s + 11400}{s^2 + 156s + 12504} \tag{8.29}$$

$$W_2(s) = 0.8475 \frac{0.1055 s^2 + 0.037 s + 0.0094}{s(s + 0.0020)^2} \tag{8.30}$$

Each of the controllers were reduced to 5th order by balanced truncation without significantly affecting the frequency response. The gain of the controllers (but not the controller structure) were scaled slightly to produce a damping ratio which ensured settling of oscillations in 10-12 seconds.

8.7.3 Performance evaluation

The eigen-values of the closed-loop system were carried out considering sequential loop closure. Table 8.4 shows the eigen-values considering only the controller K_1 for TCSC. The damping of mode #1, shown in boldface, is improved primarily with very little effect on modes #2, #3 and #4. Similarly Table 8.5 shows that the controller for SVC primarily improves the damping of mode #2, shown in boldface, besides improving mode #1 slightly. The

Table 8.4. Damping ratios and frequencies of inter-area modes with the controller for TCSC (Control loops for SVC and TCPAR open)

Mode no.	Open-loop		Closed-loop	
	ζ	$f(Hz)$	ζ	$f(Hz)$
1	**0.0626**	0.3945	**0.1544**	0.3434
2	0.0434	0.5105	0.0545	0.4991
3	0.0560	0.6269	0.0656	0.6191
4	0.0499	0.7923	0.0502	0.7918

Table 8.5. Damping ratios and frequencies of inter-area modes with the controllers for TCSC and SVC (Control loop for TCPAR open)

Mode no.	Open-loop		Closed-loop	
	ζ	$f(Hz)$	ζ	$f(Hz)$
1	0.1544	0.3434	0.1795	0.3158
2	**0.0545**	0.4991	**0.1031**	0.4549
3	0.0656	0.6191	0.0643	0.6184
4	0.0502	0.7918	0.0603	0.7864

Table 8.6. Damping ratios and frequencies of inter-area modes with the controllers for TCSC, SVC and TCPAR (All the control loops closed)

Mode no.	Open-loop		Closed-loop	
	ζ	$f(Hz)$	ζ	$f(Hz)$
1	0.1795	0.3158	0.3140	0.2682
2	0.1031	0.4549	0.2266	0.4444
3	**0.0643**	0.6184	**0.1105**	0.4585
4	0.0603	0.7864	0.0600	0.7858

controller for TCPAR primarily improves the damping of mode #3, shown in boldface, besides adding to the damping ratios of modes #1 and #2 as evidenced in Table 8.6. The combined action of the three controllers improves the damping of all three critical inter-area modes to adequate level.

8.7.4 Simulation results

To validate the robustness of the designed controllers, non-linear simulation was carried out under the same set of operating conditions described in Section 8.6.4. A three-phase solid fault for 80 ms (5 cycles) was considered as the disturbance followed by the contingency conditions depicted in the respective plots.

The angular separation between machines G1 and G15 located in different areas is shown in Fig. 8.12 under different operating scenarios. In each case, the designed controllers for TCSC, SVC and TCPAR is able to settle the oscillations within 12-15 s. The outputs of the individual FACTS devices are shown in Figs. 8.13, 8.14 and 8.15 for the same operating conditions. Appropriate limits were imposed on the variation of the control variables as seen in these figures. The limit imposed on the TCSC is the same as before. For the SVC, the output variation limit was set to -150 (inductive) to 200 (capacitive) MVAr. The limit on the phase angle of the TCPAr was set to 0 to 20 degrees.

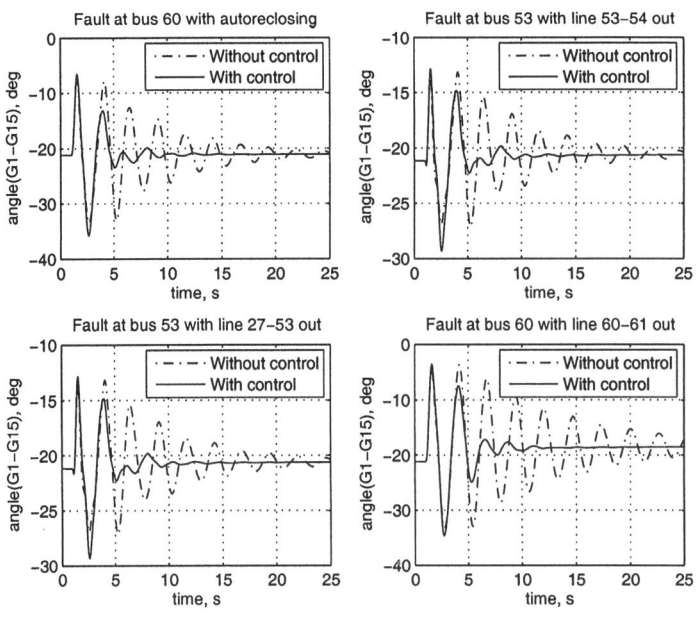

Figure 8.12. Dynamic response of the system

8.8 Summary

In this chapter, the basic concept of mixed-sensitivity design formulation has been elaborated. The problem has been translated into a generalized $\mathcal{H}_\infty$

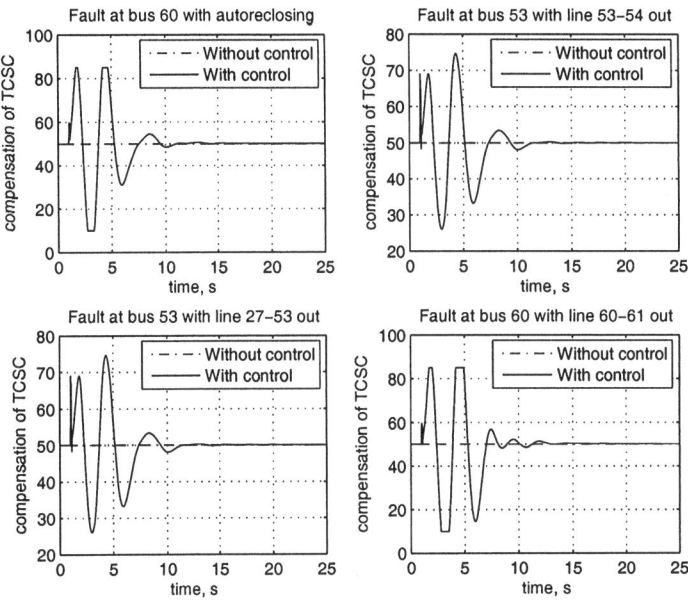

Figure 8.13. Percentage compensation of the TCSC

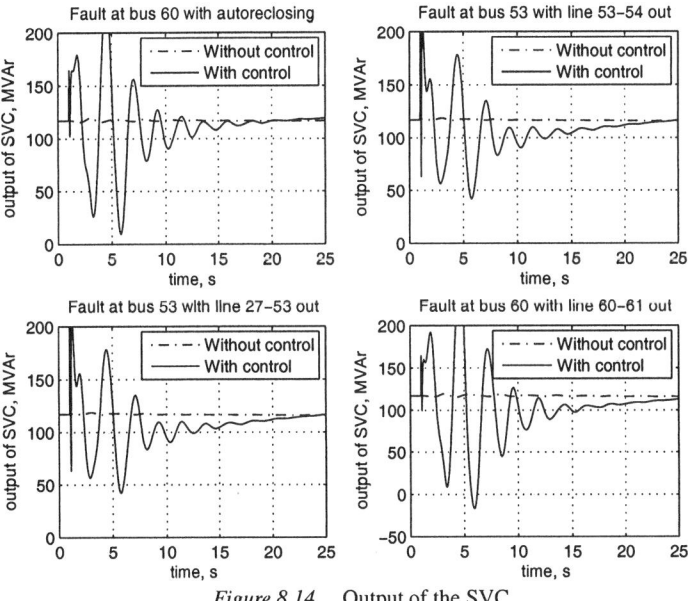

Figure 8.14. Output of the SVC

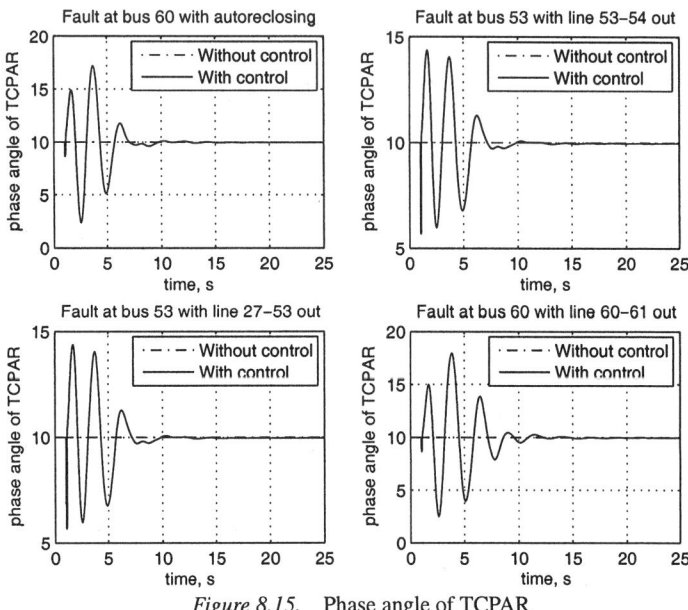

Figure 8.15. Phase angle of TCPAR

problem. The solution to the problem was sought numerically using LMIs with pole-placement. The control design methodology was illustrated by two case studies. In the first case, a centralized controller was designed for a single FACTS device. Feedback signals from three remote locations were employed for the controller. In the second case, a sequential design methodology was adopted for multiple FACTS devices. Local feedback signals were used to design decentralized controllers for individual FACTS devices. The performance and robustness of the design was validated using frequency domain analysis and non-linear simulations.

References

[mat, 1998] (1998). *Matlab Users Guide*. The Math Works Inc., USA.

[sim, 2002] (2002). *Using Simulink*. The Math Works Inc., USA.

[Boukarim et al., 2000] Boukarim, G.E., Wang, S., Chow, J.H., Taranto, G.N., and Martins, N. (2000). A comparison of classical, robust and decentralized control designs for multiple power system stabilizer. *IEEE Transactions on Power Systems*, 15(4):1287–1292.

[Chaudhuri and Pal, 2004] Chaudhuri, B. and Pal, B.C. (2004). Robust damping of multiple swing modes employing global stabilizing signals with a TCSC. *IEEE Transactions on Power Systems*, 19(1):499–506.

[Chaudhuri et al., 2003] Chaudhuri, B., Pal, B.C., Zolotas, A. C., Jaimoukha, I. M., and Green, T. C. (2003). Mixed-sensitivity approach to H_∞ control of power system oscillations employing multiple facts devices. *IEEE Transactions on Power Systems*, 18(3):1149–1156.

[Chilali and Gahinet, 1997] Chilali, M. and Gahinet, P. (1997). Multi-objective output feedback control via LMI optimization. *IEEE Transactions on Automatic Control*, 42(7):896–911.

[Gahinet and Apkarian, 1994] Gahinet, P. and Apkarian, P. (1994). A linear matrix inequality approach to H_∞ control. *International Journal of Robust and Non-linear Control*, 4(4):421–448.

[Gahinet et al., 1995] Gahinet, P., Nemirovski, A., Laub, A.J., and Chilali, M. (1995). *LMI Control Toolbox for use with Matlab*. The Math Works Inc, USA.

[Kamwa et al., 2000] Kamwa, I., Trudel, G., and Gérin-Lajoie, L. (2000). Robust design and coordination of multiple damping controllers using non-linear constrained optimization. *IEEE Transactions on Power Systems*, 15(3):1084–1092.

[Klein et al., 1995] Klein, M., Le, L.X., Rogers, G.J., Farrokpay, S., and Balu, N.J. (1995). H_∞ damping controller design in large power system. *IEEE Transactions on Power Systems*, 10(1):158–166.

[Pal et al., 1999] Pal, B.C., Coonick, A.H., and Cory, B.J. (1999). Robust damping of interarea oscillations in power systems with superconducting magnetic energy storage devices. *IEE Proceedings on Generation Transmission and Distribution*, 146(06):633–639.

[Pal et al., 2001] Pal, B.C., Coonick, A.H., and Cory, B.J. (2001). Linear matrix inequality versus root-locus approach for damping inter-area oscillations in power systems. *International Journal on Electrical Power Energy Systems*, 23(06):481–489.

[Pal et al., 2000] Pal, B.C., Coonick, A.H., Jaimoukha, I.M., and Zobaidi, H. (2000). A linear matrix inequality approach to robust damping control design in power systems with superconducting magnetic energy storage device. *IEEE Transaction on Power Systems*, 15(1):356–362.

[Paserba, 1996] Paserba, J. (1996). Analysis and control of power system oscillation. *CIGRE Special Publication 38.01.07*, Technical Brochure 111.

[Rao and Sen, 2000] Rao, P.S. and Sen, I. (2000). Robust pole placement stabilizer design using linear matrix inequalities. *IEEE Transactions on Power Systems*, 15(1):313–319.

[Scherer et al., 1996] Scherer, C., Gahinet, P., and Chilali, M. (1996). H_∞ design with pole placement constraints : An LMI approach. *IEEE Transactions on Automatic Control*, 41(3):358–367.

[Sefton and Glover, 1990] Sefton, J. and Glover, K. (1990). Pole/zero cancellations in the general H_∞ problem with reference to a two block design. *Systems and Control Letters*, 14:295–306.

[Skogestad and Postlethwaite, 2001] Skogestad, S. and Postlethwaite, I. (2001). *Multivariable Feedback Control*. John Wiley and Sons, UK.

[Taranto and Chow, 1995] Taranto, G.N. and Chow, J.H. (1995). A robust frequency domain optimization technique for tuning series compensation damping controllers. *IEEE Transactions on Power Systems*, 10(3):1219–1225.

[Taranto et al., 1998] Taranto, G.N., Wang, S., Chow, J.H., and Martins, N. (1998). Decentralized design of power system damping controllers using a linear matrix inequality algorithm. *Proceedings of VI SE-POPE*.

[Zhao and Jiang, 1995] Zhao, Q. and Jiang, J. (1995). Robust SVC controller design for improving power system damping. *IEEE Transactions on Power Systems*, 10(4):1927–1932.

[Zhou et al., 1995] Zhou, K., Doyle, J., and Glover, K. (1995). *Robust and Optimal Control*. Prentice Hall, USA.

Chapter 9

NORMALIZED $\mathcal{H}_\infty$ LOOP-SHAPING USING LINEAR MATRIX INEQUALITIES

9.1 Introduction

Over the last decade, researchers have investigated the use of $\mathcal{H}_\infty$ optimization [Klein et al., 1995, Taranto and Chow, 1995, Zhao and Jiang, 1995] and μ-synthesis [Djukanovic et al., 1999, Chen and Malik, 1995] for power system damping control design. The resulting controller has the ability to maintain stability and achieve desired performance while being insensitive to perturbations. A mixed-sensitivity design formulation with linear matrix inequality (LMI) based solution was illustrated in the previous chapter. In that approach, the designer specifies the performance requirements in terms of the weighted closed-loop transfer functions and a stabilizing controller is obtained which satisfies these criteria. One of the difficulties is that appropriate selection of the mixed-sensitivity weights is not straightforward. Moreover, it is possible for the closed-loop specifications to be made without considering the properties of the nominal system, which can often be undesirable.

A loop-shaping design methodology, however, does not suffer from the above drawbacks. It combines the characteristics of both classical open-loop shaping and $\mathcal{H}_\infty$ optimization. Zhu et. al. [Zhu et al., 2003] and Farsangi et. al. [Farsangi et al., 2002] have applied this technique for power system damping control design. However, the problem was solved analytically using a standard normalized coprime factorization approach, wherein time domain specifications in terms of minimum damping ratios (pole-placement) could not be considered explicitly in the design stage.

In this chapter, the problem of robust stabilization of a normalized coprime factor system description is converted into a generalized $\mathcal{H}_\infty$ problem. The general methodology to $\mathcal{H}_\infty$ sub-optimal solution using linear matrix inequalities (LMIs), proposed in [Gahinet et al., 1995, Chilali and Gahinet, 1997, Scherer et al., 1996], is used to obtain the solution with additional pole-placement constraints. In addition to robust stabilization of the shaped system model, a minimum damping ratio has thus been guaranteed for the critical inter-area modes of the closed-loop system.

9.2 Design approach

The normalized coprime factorization approach for loop-shaping design was proposed by McFarlane and Glover in a series of research papers [McFarlane and Glover, 1990, McFarlane and Glover, 1992, Glover and McFarlane, 1989, McFarlane and Glover, 1988]. This is a two stage design procedure that combines $\mathcal{H}_\infty$ robust stabilization with classical loop-shaping. First, the open-loop system model is augmented by pre and post-compensators to give a desired shape to the open-loop frequency response. Then the resulting shaped system model is robustly stabilized with respect to coprime factor uncertainties by solving the $\mathcal{H}_\infty$ optimization problem. In this chapter, the standard normalized coprime factorization based problem is converted into a generalized $\mathcal{H}_\infty$ problem in the LMI framework with additional pole-placement constraints [Chilali and Gahinet, 1997, Scherer et al., 1996].

9.2.1 Loop-shaping

The basic principle of $\mathcal{H}_\infty$ loop-shaping design is to pre and post-compensate the system model to shape the open-loop frequency response. The idea is to specify the performance requirements prior to robust stabilization [McFarlane and Glover, 1992]. If W_1 and W_2 are the pre and post-compensators respectively, then the shaped system model G_s is given by $G_s = W_2 G W_1$ as shown in Fig. 9.1.

The controller K is designed by solving the robust stabilization problem for the shaped system model G_s, as described later in Section 9.2.2. The equivalent feedback controller for the original system model G is obtained by augmenting the designed controller K with the compensators i.e. $K_{eq} = W_1 K W_2$ as shown in Fig. 9.1.

The primary task in loop-shaping design is to choose appropriate pre and post-compensators. Based on the recommendations in [Hyde and Glover, 1993], the following guidelines are normally used for shaping the open-loop system model [Skogestad and Postlethwaite, 2001]:

- The system inputs and outputs are properly scaled to improve conditioning of the design problem.

Normalized $\mathcal{H}_\infty$ Loop-shaping Using LMI

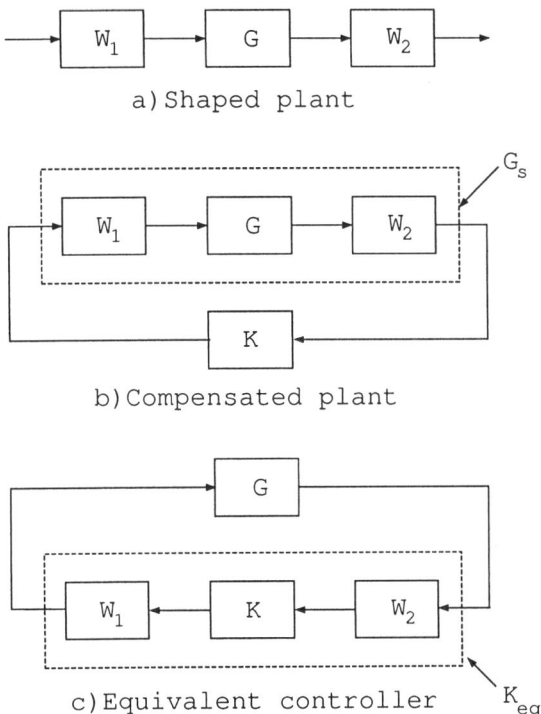

Figure 9.1. Loop-shaping design procedure

- The compensators are chosen in such a way that the singular values of the shaped system model are desirable. This would normally correspond to high gain at low frequencies, roll-off rates of approximately 20 dB/decade at desired bandwidth(s), with higher rates at high frequencies [Skogestad and Postlethwaite, 2001].

- Integral action is added at low frequencies.

It should, however, be noted that the procedure is specific to the particular application and some trial and error is involved. The maximum stability margin $\varepsilon_{\max}$, see (9.3) in Section 9.2.2, provides an indication as to whether the choice of the compensators is appropriate or not. If the margin is too small, $\varepsilon_{max} < 0.25$, then the compensators need to be modified following the above guidelines. When $\varepsilon_{max} > 0.25$, the choice is considered to be acceptable.

9.2.2 Robust stabilization

The robust stabilization of a system model is described in terms of its normalized coprime factors. A normalized left coprime factorization of a system model $G(s)$ is defined as follows:

$$G(s) = M^{-1}(s)N(s) \qquad (9.1)$$

such that the following condition is satisfied

$$MM^* + NN^* = I \qquad (9.2)$$

where, $M^*(s) = M^T(-s)$. Let us assume that the nominal open-loop shaped system model G_o can be be factorized into its left normalized coprime factors N and M^{-1}, as shown in Fig. 9.2. The perturbations around nominal N and M^{-1} are represented as Δ_N and Δ_M, respectively.

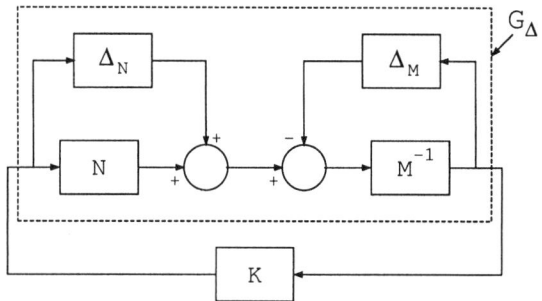

Figure 9.2. Normalized coprime factor robust stabilization problem

The largest positive number, $\varepsilon(= \varepsilon_{\max})$, such that the perturbed system model $G_\Delta = (M + \Delta_M)^{-1}(N + \Delta_N)$ can be stabilized by a controller, K, for all $\Delta = [\Delta_N, \Delta_M]; \|\Delta\|_\infty < \varepsilon$ is given by

$$\varepsilon_{\max} = \left(\inf_K \left\| \begin{bmatrix} I \\ K \end{bmatrix} (I - GK)^{-1} M^{-1} \right\|_\infty \right)^{-1} \qquad (9.3)$$

where K is chosen over the set of all stabilizing controllers [McFarlane and Glover, 1990].

The objective of the robust stabilization problem is to ensure stability under uncertainties in the system model. The larger the uncertainty against which the controller is able to ensure stability, the better is the design. In other words, obtaining a robust stabilizing controller is equivalent to maximizing the uncertainty measure ε. Therefore, the control design problem boils down to minimizing the cost function

$$\min_{K \in \mathbb{S}} \left\| \begin{bmatrix} I \\ K \end{bmatrix} (I - GK)^{-1} M^{-1} \right\|_\infty \tag{9.4}$$

The following manipulation shows that the objective function (9.4) is equivalent to a four block problem. Noting the fact that the $\mathcal{H}_\infty$ norm is invariant under right multiplication by a co-inner function [Glover and McFarlane, 1989], introduction of $[M\ N]$ in (9.4) does not affect the overall infinity norm. Therefore,

$$\left\| \begin{bmatrix} I \\ K \end{bmatrix} (I - GK)^{-1} M^{-1} \right\|_\infty$$
$$= \left\| \begin{bmatrix} I \\ K \end{bmatrix} (I - GK)^{-1} M^{-1} \begin{bmatrix} M & N \end{bmatrix} \right\|_\infty$$
$$= \left\| \begin{bmatrix} I \\ K \end{bmatrix} (I - GK)^{-1} \begin{bmatrix} I & G \end{bmatrix} \right\|_\infty$$
$$= \left\| \begin{bmatrix} S & SG \\ KS & KSG \end{bmatrix} \right\|_\infty$$

where, $S = (I - GK)^{-1}$ is the sensitivity and $\mathbb{S}$ is the set of all stabilizing controllers. The problem of robust stabilization of standard normalized coprime factor system model description is thus translated into a generalized $\mathcal{H}_\infty$ problem which can be equivalently stated as follows:

$$\min_{K \in \mathbb{S}} \left\| \begin{bmatrix} S & SG \\ KS & KSG \end{bmatrix} \right\|_\infty \tag{9.5}$$

The closed-loop transfer functions in (9.5) corresponds to robustness against specific system model/controller perturbations as mentioned below.
S: parametric perturbation of the system model
SG: additive perturbation of the controller
KS: additive perturbation of the system model
KSG: input multiplicative perturbation of the system model
Therefore, minimizing (9.5) maximizes the amount of allowable perturbations with guaranteed stability [Green and Limebeer, 1995].

The generalized regulator P for minimizing the infinity norm of the closed-loop quantities in (9.5) is given by:

$$P = \left[\begin{array}{c||c|c|c} A & B & 0 & B \\ \hline C & 0 & I & 0 \\ 0 & 0 & 0 & I \\ C & 0 & I & 0 \end{array}\right] \qquad (9.6)$$

The state representation corresponding to the generalized regulator P is given by:

$$\left[\begin{array}{c} \dot{x} \\ z \\ y \end{array}\right] = \left[\begin{array}{c||c|c} A & B_1 & B_2 \\ \hline C_1 & D_{11} & D_{12} \\ C_2 & D_{21} & 0 \end{array}\right] \left[\begin{array}{c} x \\ w \\ u \end{array}\right] \qquad (9.7)$$

where $B_1 = [B\ 0], B_2 = [B], C_1 = [C\ 0]^T, D_{11} = \begin{bmatrix} 0 & I \\ 0 & 0 \end{bmatrix}, D_{12} = [0\ I]^T, D_{21} = [0\ I]$.

The controller, $K = \left[\begin{array}{c|c} A_k & B_k \\ \hline C_k & D_k \end{array}\right]$ can be obtained by solving the $\mathcal{H}_\infty$ optimization problem given in (9.5).

For analytical solution, additional constraints (e.g. pole-placement) cannot be imposed in the synthesis stage. Therefore, in this work, the solution is obtained using a LMI formulation [Chilali and Gahinet, 1997, Scherer et al., 1996] as it offers the flexibility to impose additional pole-placement constraints which directly addresses the damping improvement issue. Having formulated the generalized regulator problem, the solution procedure is exactly the same as outlined in the previous chapter and is therefore, not repeated here.

9.3 Case study

In this section, the prototype power system model, described in Chapter 4, is considered to illustrate the control design methodology in detail. The performance and robustness of the controller is also validated.

9.3.1 Loop-shaping

To facilitate control design and to reduce the complexity of the designed controller, the nominal system model was simplified to a 7th order equivalent as described in Chapter 4. Prior to solving the $\mathcal{H}_\infty$ problem, the simplified open-loop system has to be shaped following the recommendations in Section 9.2.1. A pre-compensator was used to introduce an integral action in the low frequency region and also to reduce the overall gain of the system in order to suit the desired performance requirements. The transfer function and the frequency response of the pre-compensator is given below:

$$W_1(s) = \frac{0.106s + 0.1096}{s + 0.001} \qquad (9.8)$$

Normalized $\mathcal{H}_\infty$ Loop-shaping Using LMI

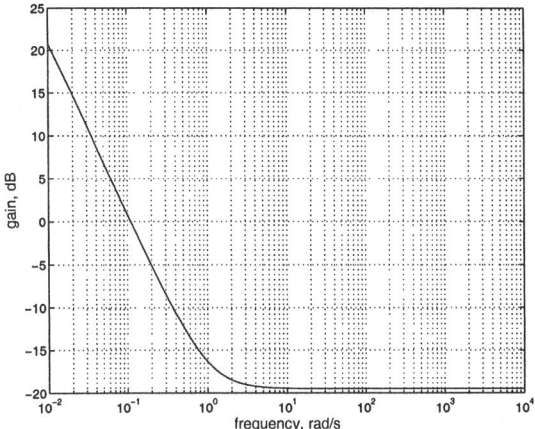

Figure 9.3. Frequency response of the pre-compensator

The three output channels were scaled with appropriate static weighting to improve the conditioning of the open loop plant. Scale factors of 1.0, 2.0, and 0.6 were used for the 1st, 2nd and 3rd outputs, respectively resulting in a post-compensator W_2 of the form:

$$W_2 = \begin{bmatrix} 1.0 & 0 & 0 \\ 0 & 2.0 & 0 \\ 0 & 0 & 0.6 \end{bmatrix} \quad (9.9)$$

The frequency response of the resulting shaped system model is shown in Fig. 9.4.

9.3.2 Control Design

The matrices A, B, C, D of shaped system model are used to formulate the generalized regulator P, defined by (9.6). Once the generalized regulator is formulated, the controller synthesis procedure is similar to that outlined in the previous chapter. The *hinfmix* function available in the *LMI Control Toolbox* [Gahinet et al., 1995] was used to perform the necessary computations. The pole-placement constraint was specified in terms of a conic sector with apex at the origin and an inner angle $\theta = 2cos^{-1}(0.15)$ as before, which ensures a minimum damping ratio of 0.15. The design converged to an optimum $\mathcal{H}_\infty$ performance index γ_{opt} of 4.873. The damping ratio and frequency of oscillation of the three critical inter-area modes is shown in Table 9.1. The damping ratio of the 2nd and 3rd modes are below 0.15 although the design was done to ensure a minimum damping ratio of 0.15. This difference is expected as the design was based on the simplified model whereas, the results shown in Table 9.1 are based on the simplified controller and full system model. The closed-loop damping

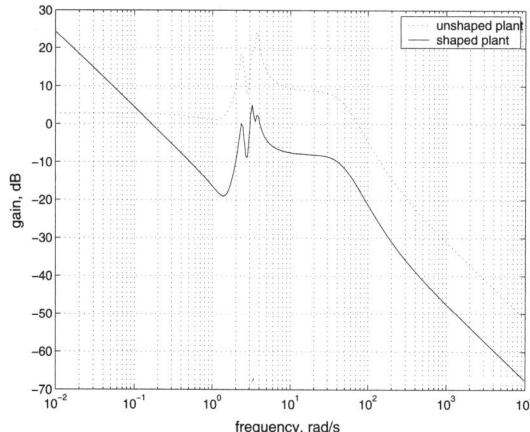

Figure 9.4. Frequency response of the reduced order original and shaped system

Table 9.1. Damping ratios and frequencies of inter-area modes with

Mode no.	Open-loop ζ	$f(Hz)$	Closed-loop ζ	$f(Hz)$
1	0.0626	0.3913	0.1701	0.3905
2	0.0435	0.5080	0.1415	0.4931
3	0.0554	0.6232	0.1152	0.6345

ratios ensure satisfactory settling of inter-area oscillations within 12-15 s as shown in the next section.

9.3.3 Simulation results

To evaluate the performance and robustness of the designed controller simulations were carried out corresponding to the same set of probable disturbance scenarios in the NETS and NYPS inter-connection, as described in the earlier chapter. The designed controller is expected to settle the inter-area oscillations within 12-15 s (performance criteria) following the disturbances. Moreover, it should be able to achieve this following any of the above disturbances (robustness) although the design is based on a nominal operating condition (no outage).

Simulations were carried out in Matlab *Simulink* [sim, 2002] for 25 s employing the *trapezoidal integration* method with a variable step size. The disturbance was created 1 s after the start of the simulation. The dynamic responses of the

Normalized $\mathcal{H}_\infty$ Loop-shaping Using LMI

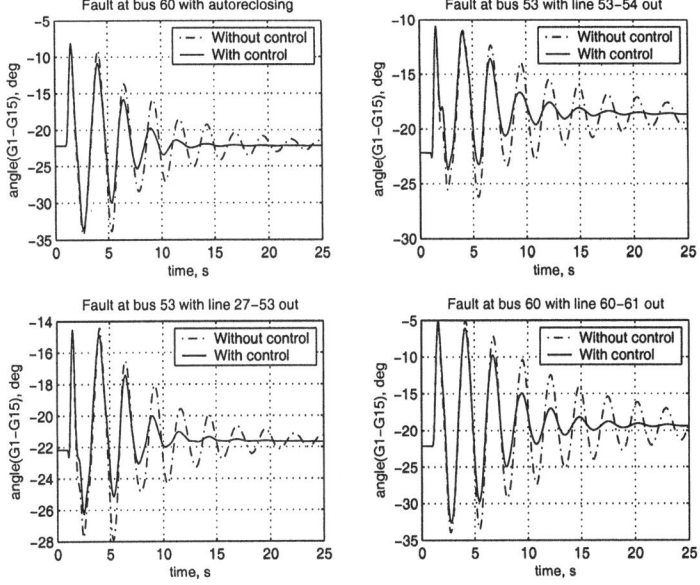

Figure 9.5. Dynamic response of the system

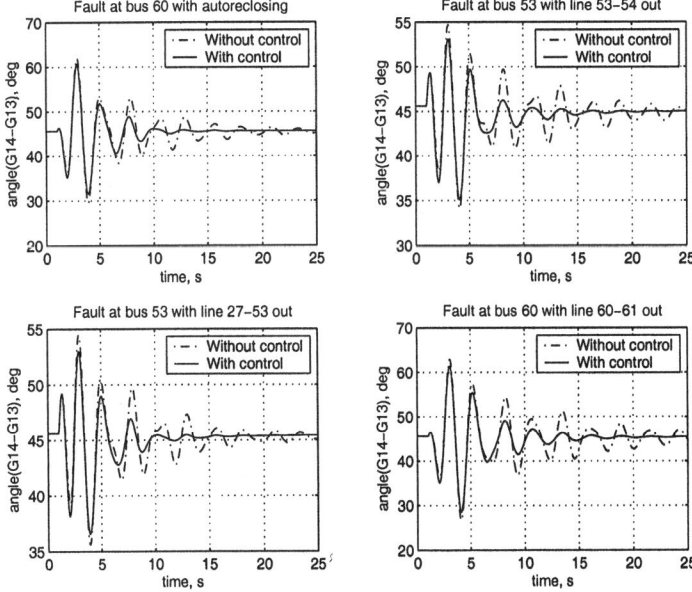

Figure 9.6. Dynamic response of the system

system following the disturbance are shown in Figs. 9.5, and 9.6.

These figures exhibit the relative angular separation between the generators located in separate geographical regions. Inter-area oscillations are mostly manifested in these angular differences and are therefore, chosen for display. It can be seen that the inter-area oscillation settles within the desired performance specification of 15 s for a range of post-fault operating conditions and thus adheres to the robustness requirement as well. A hard limit of 0.1 to 0.8 was imposed on the variation of the percentage compensation of the TCSC which is depicted in Fig. 9.7.

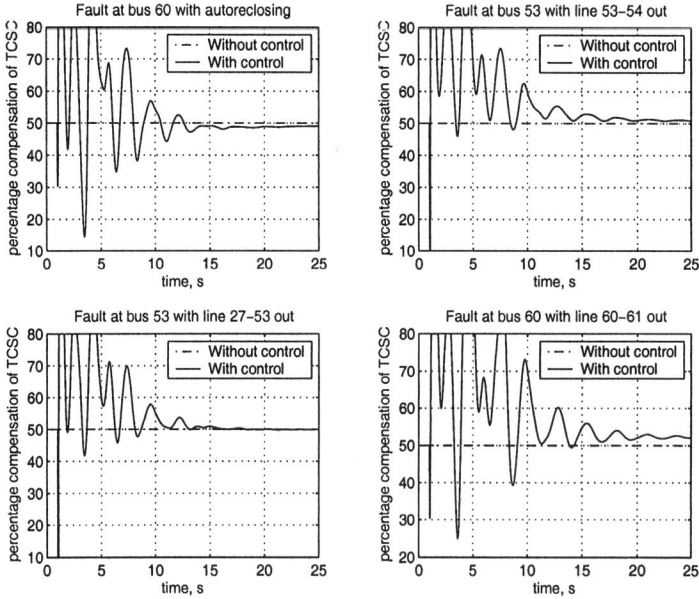

Figure 9.7. Dynamic response of the system

9.4 Summary

This chapter has demonstrated the application of the normalized $\mathcal{H}_\infty$ loop-shaping technique for design and simplification of damping controllers in the LMI framework. The first step in this design approach is to pre and post-compensate the linearized model of the power system using McFarlane and Glover loop shaping technique. The problem of robust stabilization of a normalized coprime factor system model description was translated to a generalized $\mathcal{H}_\infty$ problem. The solution was sought numerically using LMIs with additional pole-placement constraints. By imposing the constraints, a minimum damping ratio could be ensured for the critical inter-area modes which resulted in settling of oscillations within the specified time.

References

[sim, 2002] (2002). *Using Simulink*. The Math Works Inc., USA.

[Chen and Malik, 1995] Chen, S. and Malik, O.P. (1995). Power system stabilizer design using μ synthesis. *IEEE Transaction on Energy Conversion*, 10(1):175–181.

[Chilali and Gahinet, 1997] Chilali, M. and Gahinet, P. (1997). Multi-objective output feedback control via LMI optimization. *IEEE Transactions on Automatic Control*, 42(7):896–911.

[Djukanovic et al., 1999] Djukanovic, M., Khammash, M., and Vittal, V. (1999). Sequential synthesis of structured singular value based decentralized controllers in power systems. *IEEE Transaction on Power Systems*, 14(2):635–641.

[Farsangi et al., 2002] Farsangi, M.M., Song, Y.H., Fang, W.L., and Wang, X.F. (2002). Robust facts control design using H_∞ loop-shaping method. *IEE Proceedings on Generation Transmission and Distribution*, 149(3):352–357.

[Gahinet et al., 1995] Gahinet, P., Nemirovski, A., Laub, A.J., and Chilali, M. (1995). *LMI Control Toolbox for use with Matlab*. The Math Works Inc, USA.

[Glover and McFarlane, 1989] Glover, K. and McFarlane, D. (1989). Robust stabilization of normalized coprime factor plant descriptions with H_∞-bounded uncertainty. *IEEE Transactions on Automatic Control*, 34(8):821–830.

[Green and Limebeer, 1995] Green, Michael and Limebeer, David J. N. (1995). *Linear robust control*. Prentice-Hall International, Englewood Cliffs London.

[Hyde and Glover, 1993] Hyde, R.A. and Glover, K. (1993). The application of scheduled H_∞ controllers to a VSTOL aircraft. *IEEE Transactions on Automatic Control*, 38(7):1021–1039.

[Klein et al., 1995] Klein, M., Le, L.X., Rogers, G.J., Farrokpay, S., and Balu, N.J. (1995). H_∞ damping controller design in large power system. *IEEE Transactions on Power Systems*, 10(1):158–166.

[McFarlane and Glover, 1988] McFarlane, D. and Glover, K. (1988). A H_∞ design procedure using robust stabilization of normalized coprime factors. In *Proceedings of the 27th Conference on Decision and Control*, pages 1343–1348, Austin, Texas.

[McFarlane and Glover, 1992] McFarlane, D. and Glover, K. (1992). A loop shaping design procedure using H_∞ synthesis. *IEEE Transactions on Automatic Control*, 37(6):759–769.

[McFarlane and Glover, 1990] McFarlane, D.C. and Glover, K. (1990). *Robust Controller Design Using Normalized Coprime Factor Plant Descriptions*. Lecture Notes in Control and Information Sciences. Springer-Verlag, Berlin, Germany.

[Scherer et al., 1996] Scherer, C., Gahinet, P., and Chilali, M. (1996). H_∞ design with pole placement constraints : An LMI approach. *IEEE Transactions on Automatic Control*, 41(3):358–367.

[Skogestad and Postlethwaite, 2001] Skogestad, S. and Postlethwaite, I. (2001). *Multivariable Feedback Control*. John Wiley and Sons, UK.

[Taranto and Chow, 1995] Taranto, G.N. and Chow, J.H. (1995). A robust frequency domain optimization technique for tuning series compensation damping controllers. *IEEE Transactions on Power Systems*, 10(3):1219–1225.

[Zhao and Jiang, 1995] Zhao, Q. and Jiang, J. (1995). Robust SVC controller design for improving power system damping. *IEEE Transactions on Power Systems*, 10(4):1927–1932.

[Zhu et al., 2003] Zhu, Chuanjiang, Khammash, Mustafa, Vittal, Vijay, and Qiu, Wenzheng (2003). Robust power system stabilizer design using H_∞ loop shaping approach. *IEEE Transactions on Power Systems*, 18(2):810–818.

Chapter 10

$\mathcal{H}_\infty$ CONTROL FOR TIME-DELAYED SYSTEMS

10.1 Introduction

In large scale power systems, inter-area response might me more effectively damped through the use of wide-area measurements system (WAMS). Fibre optic communication is a promising technology for sensing and measurement systems used to implement the wide-area measurement systems. The advent of global positioning system (GPS) technology has made time stamping fairly routine and measurements of phase and other temporal information is, therefore, attainable through the use of commercially available equipment [Heydt et al., 2001]. The information architecture, proposed in [Xie et al., 2002], is capable of providing timely, secure, reliable information exchange among various entities in a power system.

With the rapid advancements in wide-area measurement systems technology coupled with a fast and reliable data transmission infrastructure, the prospect of centralized control of power systems has gained momentum. Damping of inter-area oscillations through remote measurements has therefore, become realistic from the implementation point of view. From an economic viewpoint, implementation of centralized control using remote signals often turns out to be more cost effective than installing new control devices [Chow et al., 2000]. The obvious question is at what speed are these remote measurements available to the control site. Employing phasor measurement units (PMUs), it is possible to deliver the signals at a speed as high as 30 Hz sampling rate [Kamwa et al., 2001, Heydt et al., 2001]. It is possible to deploy the PMUs at strategic locations of the grid and obtain a coherent picture of the entire network in real time [Heydt et al., 2001].

However, the infrastructure cost and associated complexities restrict the use of such sophisticated signal transmission hardware on a large commercial scale. As a more viable alternative, the existing communication channels are often used to transmit the signals from remote locations. The major problem is the delay involved between the instant of measurement and that of the signal being available to the controller. A conservative estimate of the delay can typically be in the range of 0.5-1.0 s depending on the distance, transmission protocol and several other factors. As the delay is of a sizable amount, it should be accounted for in the design stage itself to ensure effective control action. In the previous chapters, no time-delay was considered as fast transmission of the necessary signals (typically within 0.02-0.05 s) was assumed. As the delay was considerably less than the smallest time period of the inter-area modes, it was not necessary to consider it in the design stage.

However, in this chapter, the power system is treated as a dead-time system involving a long delay in transmitting the measured signals from remote locations to the controller site. It is not straightforward to control such time-delayed systems [Zhong, 2003]. The Smith predictor [Smith, 1957, Smith, 1958] approach, proposed in the early fifties, was the first effective tool for handling such control problems. Normal $\mathcal{H}_\infty$ controllers are unlikely to guarantee satisfactory control action for time-delayed systems. The difficulties associated with the design of $\mathcal{H}_\infty$ controllers for such systems and their potential solution using the predictor based approach is discussed in this chapter.

Since the 1950s, a number of variations of the Smith predictors have been proposed in the literature. One of the drawbacks of the classical Smith predictor (CSP) approach is that it is very difficult to ensure a minimum damping ratio of the closed loop poles when the open-loop system model has lightly damped poles (as often encountered in power systems). A modified Smith predictor (MSP) approach, proposed in [Wantanable and Ito, 1981], was used to overcome the drawbacks of the CSP. However, the control design using the MSP approach might run into numerical problems for systems with fast stable poles. To overcome the drawbacks of CSP and MSP, a unified Smith predictor (USP) approach was proposed very recently by Zhong [Zhong and Weiss, 2004]. The USP approach effectively combines the advantageous features of both the CSP and MSP.

In this chapter, we have illustrated the application of the USP approach for designing a damping controller for power system with time-delayed signals. The predictor based control design methodology is illustrated by a case study on the same power system model described in Chapter 4. The performance and robustness of the designed controller is validated using frequency domain analysis and time domain simulations.

10.2 Smith predictor for time-delayed or dead-time systems: an overview

In a time-delayed or dead-time system, either the measured output takes a certain time before it affects the control input or the action of the control input takes a certain time before it influences the measured outputs. Typical dead-time systems consist of input and/or output delays. The general control setup for a system having an output delay is shown in Fig. 10.1, where,

$$P(s) = \begin{bmatrix} P_{11}(s) & P_{12}(s) \\ P_{21}(s) & P_{22}(s) \end{bmatrix} \quad (10.1)$$

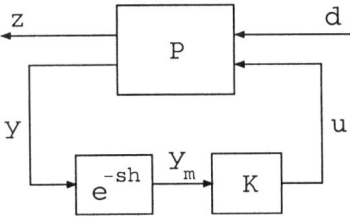

Figure 10.1. Control setup for dead-time systems

and the closed-loop transfer matrix from d to z is: $T_{zd}(s) = P_{11} + P_{12} K e^{-sh}(I - P_{22} K e^{-sh})^{-1} P_{21}$. This suggests that there exists an instantaneous response through the path P_{11} (path 1 in Fig. 10.2) without any delay. An equivalent structure is shown in Fig. 10.2.

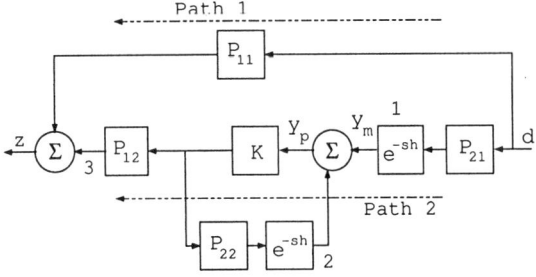

Figure 10.2. An equivalent representation of dead-time systems

It can be seen that during the period $t = 0 \sim h$ after d is applied, the output z is not controllable, since it is only determined by P_{11} and d with no response coming through the controlled path (path 2 in Fig. 10.2). This means that the $\mathcal{H}_\infty$ performance index $\|T_{zd}\|_\infty$ is likely to be dominated by a response that cannot be controlled which is not desirable. It is extremely difficult to design a controller for such systems [Zhong, 2003].

Smith predictor (SP), is the first effective tool for tackling such control problems. The primary idea is to eliminate any uncontrollable response that is likely to govern the $\mathcal{H}_\infty$ performance index. One possible way of achieving this is to introduce a uniform delay in both paths (path 1 and path 2) as shown in Fig. 10.4. There are two steps towards achieving this. Firstly, the delay blocks (e^{-sh}) at points 1 and 2 need to be shifted to point 3 by introducing a suitable predictor block in parallel with K. Secondly, a delay block needs to be introduced into path 1.

The first step is achieved by introducing a Smith predictor block $Z(s) = P_{22}(s) - P_{22}(s)e^{-sh}$, as shown by the dotted box in Fig. 10.3. The second task of bringing a delay in path 1 is done while forming the generalized regulator prior to control design. Presence of the predictor block Z and the delay in path 1 ensures that the responses (through path 1 and path 2) governing the performance index are delayed uniformly as shown in Fig. 10.4.

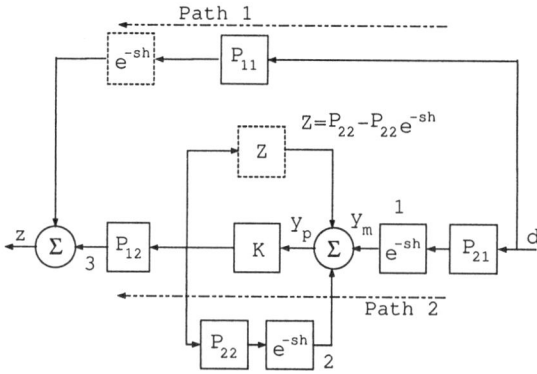

Figure 10.3. Introduction of Smith predictor and delay block

A predictor-based controller for the dead-time system $P_h(s) = P_{22}(s)e^{-sh}$ consists of a predictor $Z = P_{22} - P_{22}e^{-sh}$ and a stabilizing compensator K, as shown in Fig. 10.5. The predictor Z is an exponentially stable system such that $P_h + Z$ is rational i.e. it does not involve any uncontrollable response governing the $\mathcal{H}_\infty$ performance index.

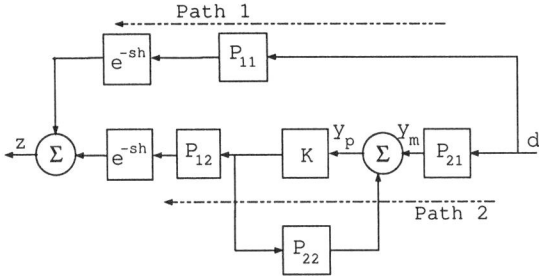

Figure 10.4. Uniform delay in both paths

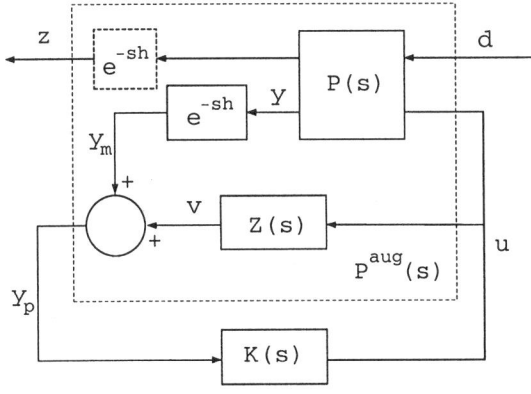

Figure 10.5. Smith predictor formulation

To overcome the shortcomings of the CSP approach for systems having poorly damped open-loop poles, the MSP approach was introduced which enabled the robust control problems of dead-time systems to be solved similarly as in finite dimensional situations [Zhong, 2003].

Let us consider a generalized delay-free system given by:

$$P(s) = \left[\begin{array}{c|cc} A & B_1 & B_2 \\ \hline C_1 & D_{11} & D_{12} \\ C_2 & D_{21} & D_{22} \end{array}\right] = \left[\begin{array}{cc} P_{11} & P_{12} \\ P_{21} & P_{22} \end{array}\right] \quad (10.2)$$

For a delay of h, the generalized regulator formulation using the MSP approach would be [Zhong, 2003]

$$\tilde{P}(s) = \left[\begin{array}{c|cc} A & e^{Ah}B_1 & B_2 \\ \hline C_1 & 0 & D_{12} \\ C_2 e^{-Ah} & D_{21} & 0 \end{array}\right] \quad (10.3)$$

where, $\tilde{P}(s)$ is $P^{aug}(s)$ including the effect of the delay block in between d and z, as shown in Fig. 10.5.

The computation of matrix exponential e^{-Ah} in (10.3) suffers from numerical problems especially for systems having fast stable eigen-values. In the worst case it might well be non-computable. This problem can even arise with reasonably small amount of delays if some of the stable eigen-values are very fast.

In $\mathcal{H}_\infty$ mixed-sensitivity formulation for power system damping control design, the presence of fast stable eigen-values in the augmented system cannot be ruled out, the possible sources being the fast sensing circuits ($T \sim 0.02$ s), fast damper circuits ($T \sim 0.05$ s) and even the weighting filters. These often lead to numerical instability while solving the problem using LMIs. These problems are overcome through the use of the USP [Zhong and Weiss, 2004] formulation, achieved by decomposing the delay free system model P into a critical part P_c and a non-critical part P_{nc}. The critical part contains the poorly damped poles of the system whereas, the non-critical part consists of poles with sufficiently large negative real values. The next section describes the generalized problem formulation using this approach.

10.3 Problem formulation using unified Smith predictor

As indicated in the previous section, the first step towards formulating the control problem using the USP approach is to decompose the delay-free system model into critical and non-critical parts. This is normally done by applying a suitable linear coordinate transformation on the state space representation of the system. In this work, a suitable transformation matrix V was chosen such that the transformed matrix $J = V^{-1}AV$ is in the Jordan canonical form and is free from complex entries. The transformation matrix V was chosen using the *'eig'* function available in Matlab [mat, 1998]. The elements of the transformed matrix J were converted from complex diagonal form to a real diagonal form using the *'cdf2rdf'* function in Matlab [mat, 1998]. The transformed augmented delay-free system model P_{22}^t is given by

$$P_{22}^t(s) = \left[\begin{array}{c|c} V^{-1}AV & V^{-1}B_2 \\ \hline C_2V & D_{22} \end{array}\right] = \left[\begin{array}{cc|c} A_c & 0 & B_c \\ 0 & A_{nc} & B_{nc} \\ \hline C_c & C_{nc} & D_{22} \end{array}\right] \quad (10.4)$$

where, A_c is the critical and A_{nc} is the non-critical part of A. The augmented system model P_{22}^t can be split as $P_{22}^t = P_c + P_{nc}$ where:

$$P_c(s) = \left[\begin{array}{c|c} A_c & B_c \\ \hline C_c & D_{22} \end{array}\right] \quad (10.5)$$

and

$$P_{nc}(s) = \left[\begin{array}{c|c} A_{nc} & B_{nc} \\ \hline C_{nc} & 0 \end{array}\right] \quad (10.6)$$

The predictor for the critical part is formulated using the MSP approach by applying a completion operator [Zhong, 2003]. On a rational transfer matrix $G(s) = D + C(sI - A)^{-1}B$, the completion operator $\pi_h\{e^{-sh}G\}$ is defined as follows:

$$\pi_h\{e^{-sh}G\} \triangleq \left[\begin{array}{c|c} A & B \\ \hline Ce^{-Ah} & 0 \end{array}\right] - \left[\begin{array}{c|c} A & B \\ \hline C & D \end{array}\right]e^{-sh} \quad (10.7)$$

Using (10.7), the predictor for the critical part P_c is given by (10.8), see [Zhong, 2003] for details.

$$Z_c(s) = \pi_h\{e^{-sh}P_c\} \triangleq \left[\begin{array}{c|c} A_c & B_c \\ \hline C_c e^{-A_c h} & 0 \end{array}\right] - \left[\begin{array}{c|c} A_c & B_c \\ \hline C_c & D_c \end{array}\right]e^{-sh}$$
$$= P_c^{aug}(s) - P_c(s)e^{-sh} \quad (10.8)$$

The predictor for the non-critical part is constructed following the CSP formulation and is given by:

$$Z_{nc}(s) = P_{nc}(s) - P_{nc}(s)e^{-sh} \quad (10.9)$$

The USP, denoted by Z, is simply the sum of Z_c and Z_{nc}, as shown in Fig. 10.6.

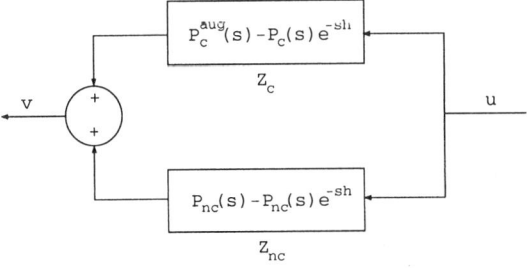

Figure 10.6. Unified Smith predictor

It is given by:

$$Z(s) = P_{22}^{aug}(s) - P_{22}^{t}(s)e^{-sh} \quad (10.10)$$

where $P_{22}^{aug} = P_{nc} + P_c^{aug}$. Using (10.4), (10.6) and (10.8), the realization for P_{22}^{aug} can be expressed in the form:

$$P_{22}^{aug} = \left[\begin{array}{c|c} A & B_2 \\ \hline C_2 E_h & 0 \end{array}\right] \quad (10.11)$$

where,

$$E_h = V \begin{bmatrix} e^{-A_c h} & 0 \\ 0 & I_{nc} \end{bmatrix} V^{-1} \quad (10.12)$$

The augmented system model P^{aug} is obtained by connecting the original dead-time system and the USP in parallel, see Fig.10.5, where the new set of measured outputs are y_p. The expression for P^{aug} is given by:

$$P^{aug} = \begin{bmatrix} P_{11}(s) & P_{12}(s)e^{-sh} \\ P_{21}(s) & P_{22}(s)^{aug} \end{bmatrix} \quad (10.13)$$

The generalized regulator $\tilde{P}$ can be formulated from P^{aug} after inserting the delay block e^{-sh} in between d and z as shown by a dotted box in Fig. 10.5. The steps for arriving at the final expression for $\tilde{P}$ is detailed in [Zhong and Weiss, 2004]. The final form of the generalized regulator is the following:

$$\tilde{P} = \left[\begin{array}{cc|cc} A & 0 & E_h^{-1} B_1 & B_2 \\ 0 & A_{nc} & \begin{bmatrix} 0 & e^{A_{nc}h} - I_{nc} \end{bmatrix} V^{-1} B_1 & 0 \\ \hline C_1 & C_1 V \begin{bmatrix} 0 \\ I_{nc} \end{bmatrix} & 0 & D_{12} \\ C_2 E_h & 0 & D_{21} & 0 \end{array}\right] \quad (10.14)$$

Having formulated the generalized regulator $\tilde{P}$ following the USP approach, the objective is to design a controller K to meet the desired performance specifications. If K ensures the desired performance for $\tilde{P}$, then the controller predictor combination $K_e = K(I - ZK)^{-1}$ is guaranteed to achieve the same for the original dead-time system [Zhong, 2003].

10.4 Case study

In this section, the prototype power system model described in Chapter 4 is considered to illustrate the control design methodology in details. The performance and robustness of the design is also validated.

In the previous chapters, it was assumed that the remote signals were available at the control site with a negligible time-delay of 0.02 s. However, in this chapter a sizeable amount of delay is considered to illustrate the effectiveness of the design technique. A conservative estimate of 0.75 s was considered as the signal transmission delay with a view to cover the worst case scenario.

10.4.1   Control design

The control design problem was formulated using the standard mixed-sensitivity approach [Chaudhuri et al., 2003, Chaudhuri and Pal, 2004] with modifications to include the effect of delay. The overall control setup is shown in Fig. 10.7.

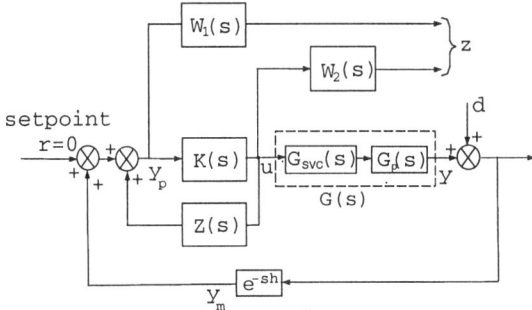

Figure 10.7. Control setup with mixed-sensitivity design formulation

where,
$G_p(s)$: power system model,
$G_{svc}(s)$: SVC model,
$W_1(s), W_2(s)$: weighting filters,
$K(s)$: controller to be designed,
$Z(s)$: Smith predictor,
d: disturbance at the system output,
z: weighted exogenous outputs,
y_m: measured output and
u: control input.

The design objective is the following:

Find a controller K from the set of internally stabilizing controllers $\mathbb{S}$ such that

$$\min_{K \in \mathbb{S}} \left\| \begin{bmatrix} W_1 S \\ W_2 K S \end{bmatrix} \right\|_\infty < 1 \qquad (10.15)$$

where $S = (I - GK)^{-1}$ is the sensitivity. The solution to the problem was sought numerically using the LMI solver with additional pole-placement constraint.

Once the generalized regulator is formulated following the steps given in Section 10.3, the basic steps for control design are exactly similar to those outlined in the previous chapter. The weights were chosen as follows.

$$W_1(s) = \frac{100}{s + 100}, \quad W_2(s) = \frac{100s}{s + 100} \quad (10.16)$$

The frequency responses for the weighting filters are shown in Fig. 10.8 and they are in accordance with the basic requirement of mixed-sensitivity design i.e. $W_1(s)$ should be a low pass filter for output disturbance rejection and $W_2(s)$ should be a high-pass filter in order to reduce the control effort and to ensure robustness against additive uncertainties in the system model in the high frequency range.

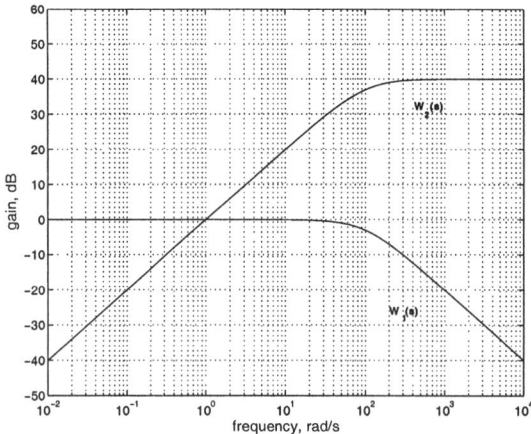

Figure 10.8. Frequency response of the weighting filters

To facilitate control design and to reduce the complexity of the designed controller, the nominal system model was reduced to a 7th order equivalent as described in Chapter 4. Using the simplified system model and the above mentioned weights, the generalized problem was formulated according to (10.14). The solution was numerically sought using suitably defined objectives in the argument of the function *hinfmix* of the *LMI Toolbox* in Matlab [mat, 1998]. The pole-placement constraint was imposed by using a 'conic sector' of inner angle $2cos^{-1}(0.175)$ with apex at the origin. The order of the controller ob-

$\mathcal{H}_\infty$ *Control for Time-delayed Systems*

tained from the LMI solution was equal to the reduced order augmented system order plus the order of the weights. The designed controller was reduced to a 8th order equivalent using Schur's method without affecting the frequency response as shown in Fig. 10.9.

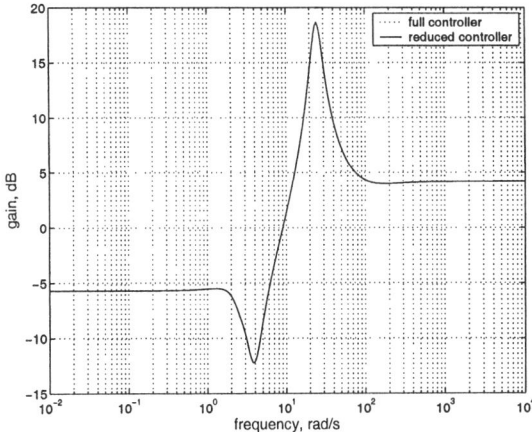

Figure 10.9. Frequency response of the full and reduced controller

10.4.2 Performance evaluation

The eigen-values of the closed-loop system were computed to examine the performance of the designed controller in terms of improving the damping ratios of the inter-area modes. A 4th order Pade approximation was used to represent the delay (= 0.75 s) in the frequency domain. The results are summarized in Tables 10.1. It can be seen that the damping ratios of the three critical inter-area modes, shown in boldface, are improved in the presence of the controller. The damping action of the controller was examined under different power flow levels and different types of loads. It is evident from the results of Tables 10.2 and 10.3 that the designed controller provides robust damping for varying operating conditions.

10.4.3 Simulation results with TCSC

A series of simulations were carried out under the same set of disturbances, as described in the previous chapter. The 'transport delay' blocks in *Simulink* [sim, 2002] were used to simulate the signal transmission delays and also to implement the USP. The dynamic responses of the system following these disturbances are shown in Fig. 10.10.

The relative angular separation between the generators G1 and G15 which are located in separate geographical regions shows that inter-area oscillations

Table 10.1. Damping ratios and frequencies of the inter-area modes

Mode no.	No control ζ	$f(Hz)$	With control ζ	$f(Hz)$
1	0.0626	0.3913	**0.1949**	0.3753
2	0.0435	0.5080	**0.1462**	0.5157
3	0.0554	0.6232	**0.1495**	0.6108
4	0.0499	0.7915	**0.0742**	0.7966

Table 10.2. Damping ratios and frequencies of the critical inter-area modes at different levels of power flow between NETS and NYPS

Power flow (MW)	Mode 1 ζ	$f(Hz)$	Mode 2 ζ	$f(Hz)$	Mode 3 ζ	$f(Hz)$
100	**0.2424**	0.3675	**0.1453**	0.5140	**0.1388**	0.6538
500	**0.2233**	0.3745	**0.1449**	0.5148	**0.1479**	0.6251
700	**0.1949**	0.3753	**0.1462**	0.5157	**0.1495**	0.6108
900	**0.1765**	0.3712	**0.1509**	0.5169	**0.1435**	0.5884

Table 10.3. Damping ratios and frequencies of the critical inter-area modes for different load models

Type of load	Mode 1 ζ	$f(Hz)$	Mode 2 ζ	$f(Hz)$	Mode 3 ζ	$f(Hz)$
CI	**0.1949**	0.3753	**0.1462**	0.5157	**0.1495**	0.6108
CC+CI	**0.1835**	0.3881	**0.1410**	0.5277	**0.1344**	0.6616
CP+CI	**0.1714**	0.3918	**0.1366**	0.5385	**0.1195**	0.6230
Dynamic	**0.1913**	0.3709	**0.1617**	0.5229	**0.1495**	0.6092

are settled within 12-15 s even though the feedback signals arrive at the control location after a finite time-delay of 0.75 s. The variation of the TCSC output is shown in Fig. 10.11.

Although the controller is designed considering a fixed delay of 0.75 s, there can always be some variation in the amount of delay that is actually encountered in practice. The designed controller has therefore been tested for a range of possible signal transmission delays from 0.5 s to 1.0 s, the controller performs

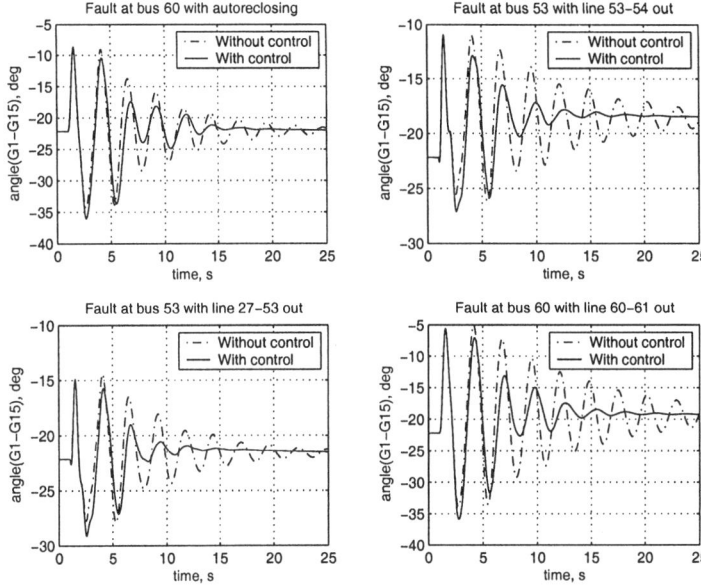

Figure 10.10. Dynamic response of the system with TCSC installed; controller designed with 0.75 s delay

satisfactorily as shown in Fig. 10.12. However, at 0.5 s and above 1.0 s delay, the performance deteriorates.

To demonstrate the drawback of the conventional $\mathcal{H}_\infty$ design with a delay-free system, a separate controller was designed for the TCSC without considering any delay at the design stage. The design was carried out following the LMI based methodology described in [Chaudhuri et al., 2003, Chaudhuri and Pal, 2004]. The controller performed satisfactorily both in time and frequency domain for delays up to 0.1 s. For larger time delays, the performance of the controller appeared to be poor and was even worse with increasing amount of delay. Simulations were carried out with this controller considering a time delay of 0.75 s. The simulation results following the same disturbance described above are shown in Figs. 10.13.

It is clear that the system performance deteriorates considerably revealing the fact that if the delay is more than the time-period corresponding to the dominant modes and it is not taken into account during the design stage the controller might not perform as expected. Therefore, it is necessary to include the delay in the control design formulation itself. The results demonstrate a

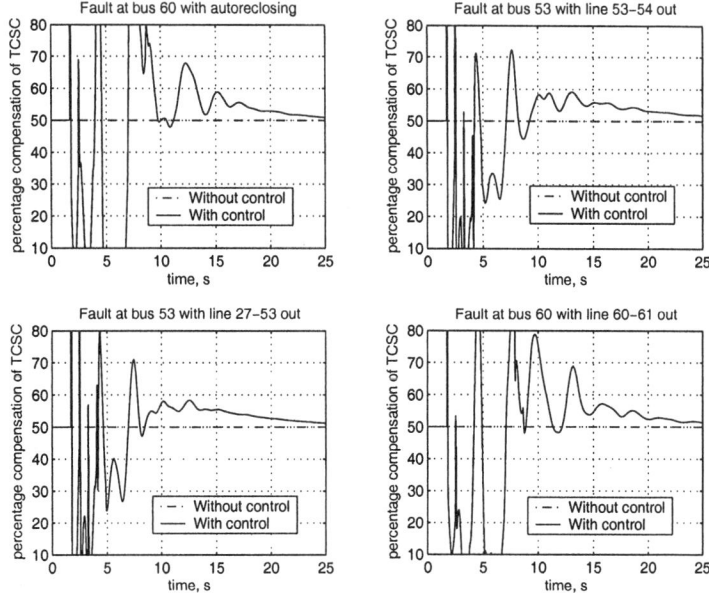

Figure 10.11. Output of the TCSC

potential application of the USP approach for power system damping control design involving a finite amount of signal transmission delay.

10.5 Simulation results with SVC

Time-domain simulation was also carried out with an SVC installed in the system.

The controller for the SVC was designed following exactly the same procedure as used for the TCSC. The same disturbance as considered in the previous section has been considered. The dynamic response of the system following this disturbance is shown in Fig. 10.14 which exhibits the relative angular separation between the generators G1 and G15. It is clear that the inter-area oscillation is damped out in 12-15 s even though the feedback signals arrive at the control location after a finite time delay of 0.75 s. The variation of the output of the SVC is shown in Fig. 10.15 which is within a range of -1.5 pu to 2.0 pu.

The performance of the controller for different delays is shown in Figs. 10.16 and 10.17. The simulation results, shown in Fig.10.18, demonstrate the detrimental effect of not considering the delay at the design stage, if there is one in practice.

$\mathcal{H}_\infty$ Control for Time-delayed Systems

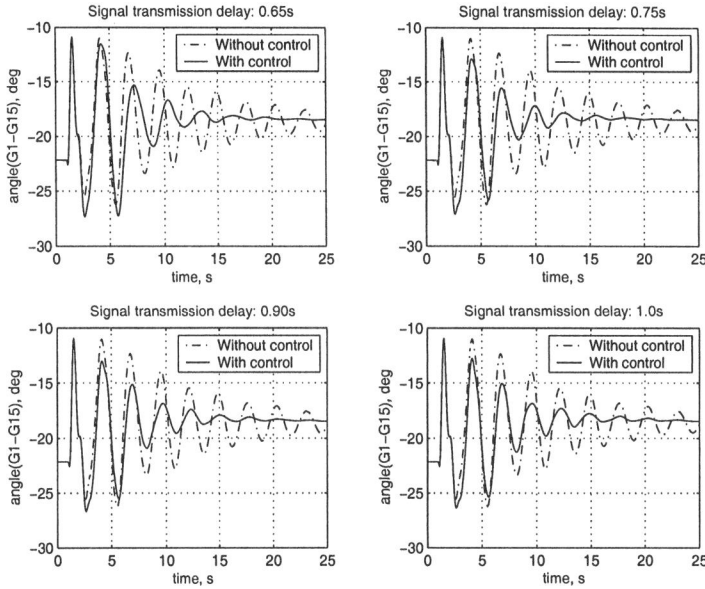

Figure 10.12. Dynamic response of the system with a delay of 0.5 s

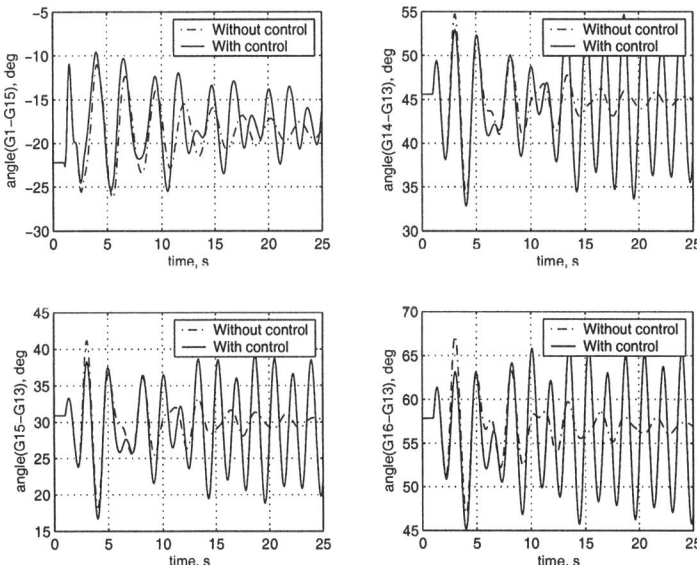

Figure 10.13. Dynamic response of the system; controller designed without considering delay

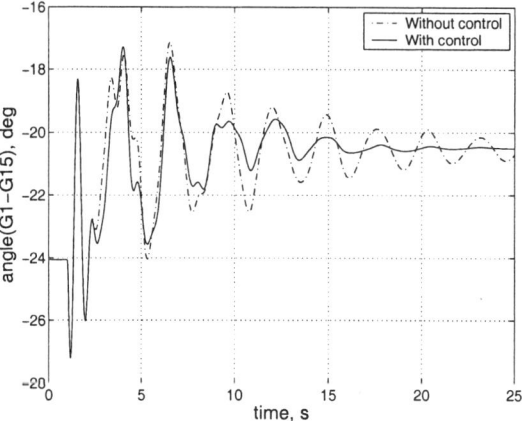

Figure 10.14. Dynamic response of the system with SVC; controller designed considering delay

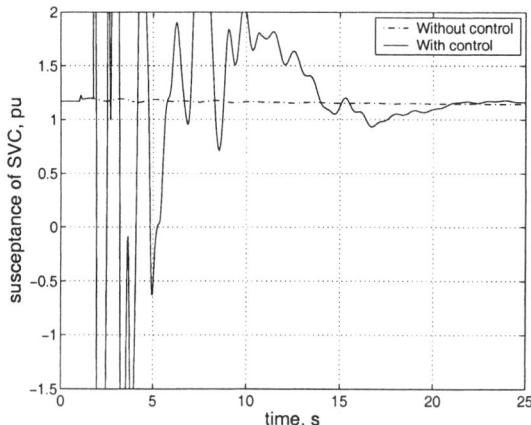

Figure 10.15. Output of the SVC

The results highlight a potential application of the USP approach for power system damping control design with different types of FACTS devices where, the transmission of remote feedback signals involves a finite amount of time-delay.

10.6 Summary

In this chapter, a methodology for power system damping control design accounting for delayed arrival of feedback signals from remote locations is described. A predictor based $\mathcal{H}_\infty$ control design strategy has been presented for

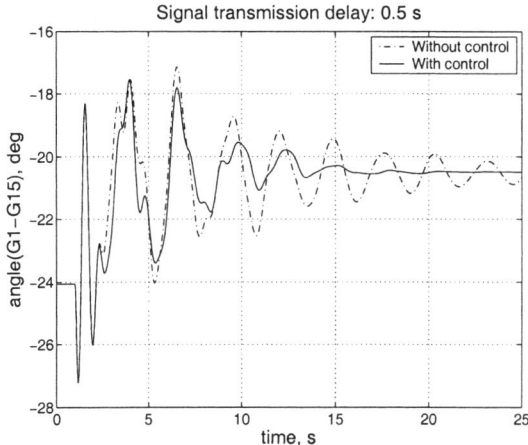

Figure 10.16. Dynamic response of the system with a delay of 0.5 s

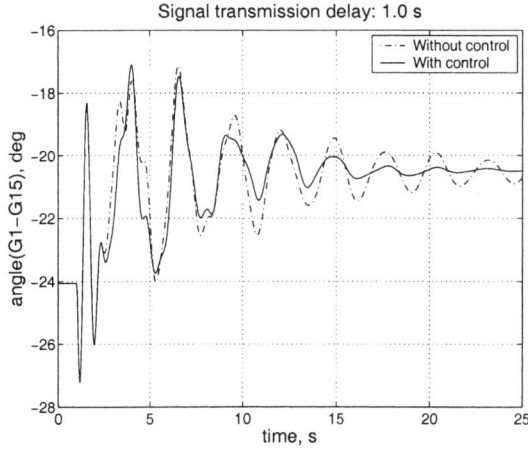

Figure 10.17. Dynamic response of the system with a delay of 1.0 s

such a time-delayed system. The design procedure based on the USP approach has been applied for centralized design of a power system damping controller through two different types of FACTS devices i.e. TCSC and SVC. A combination of the USP and the designed controller was found to work satisfactorily under different operating scenarios even though the stabilizing signals could reach the controller site only after a finite time.

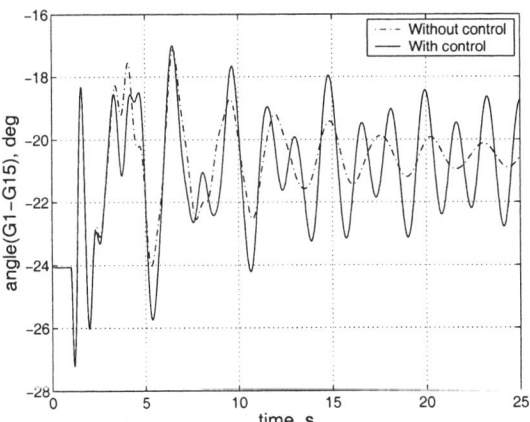

Figure 10.18. Dynamic response of the system; controller designed without considering delay

Here, a fixed time delay has been considered for all the communication channels. In practice, this might not always be the case as the distances from the measurement sites differ. Therefore, different amount of delay for each signal needs to be considered during the design. Further research is currently being continued in this area.

References

[mat, 1998] (1998). *Matlab Users Guide*. The Math Works Inc., USA.

[sim, 2002] (2002). *Using Simulink*. The Math Works Inc., USA.

[Chaudhuri and Pal, 2004] Chaudhuri, B. and Pal, B.C. (2004). Robust damping of multiple swing modes employing global stabilizing signals with a TCSC. *IEEE Transactions on Power Systems*, 19(1):499–506.

[Chaudhuri et al., 2003] Chaudhuri, B., Pal, B.C., Zolotas, A. C., Jaimoukha, I. M., and Green, T. C. (2003). Mixed-sensitivity approach to H_∞ control of power system oscillations employing multiple facts devices. *IEEE Transactions on Power Systems*, 18(3):1149–1156.

[Chow et al., 2000] Chow, J.H., Sanchez-Gasca, J.J., Ren, H., and Wang, S. (2000). Power system damping controller design using multiple input signals. *IEEE Control Systems Magazine*, 20(4):82–90.

[Heydt et al., 2001] Heydt, G.T., Liu, C.C., Phadke, A.G., and Vittal, V. (2001). Solutions for the crisis in electric power supply. *IEEE Computer Applications in Power*, 14(3):22–30.

[Kamwa et al., 2001] Kamwa, I., Grondin, R., and Hebert, Y. (2001). Wide-area measurement based stabilizing control of large power systems - a decentralized/hierarchical approach. *IEEE Transactions on Power Systems*, 16(1):136–153.

REFERENCES

[Smith, 1957] Smith, O.J.M. (1957). Closer control of loops with dead time. *Chem. Eng. Progress*, 53(5):217–219.

[Smith, 1958] Smith, O.J.M. (1958). *Feedback Control Systems*. McGraw-Hill Book Company Inc., USA.

[Wantanable and Ito, 1981] Wantanable, K. and Ito, M. (1981). A process-model control for linear systems with delay. *IEEE Transactions on Automatic Control*, 26(6):1261–1269.

[Xie et al., 2002] Xie, Zhaoxia, Manimaran, G., Vittal, Vijay, Phadke, A. G., and Centeno, Virgilio (2002). An information architecture for future power systems and its reliability analysis. *IEEE Transactions on Power Systems*, 17(3):857–863.

[Zhong, 2003] Zhong, Q. C. (2003). H_∞ control of dead time systems based on a transformation. *Automatica*, 39:361–366.

[Zhong and Weiss, 2004] Zhong, Q.-C. and Weiss, G. (2004). A unified smith predictor based on the spectral decomposition of the plant. *International Journal of Control*, 77(15):1362–1371.

Appendix A
16-machine, 5-area System Power Flow Data

Table A.1. Machine bus data

Bus number	Voltage (pu)	Power generation (pu)
1	1.0450	2.50
2	0.9800	5.45
3	0.9830	6.50
4	0.9970	6.32
5	1.0110	5.05
6	1.0500	7.00
7	1.0630	5.60
8	1.0300	5.40
9	1.0250	8.00
10	1.0100	5.00
11	1.0000	10.00
12	1.0156	13.50
13	1.0110	35.91
14	1.0000	17.85
15	1.0000	10.00
16	1.0000	40.00

Table A.2. Load bus data

Bus number	Real load (pu)	Reactive load (pu)
17	60.00	3.0000
18	24.70	1.2300
19	0	0
20	6.80	1.0300
21	2.74	1.1500
22	0	0
23	2.48	0.8500
24	3.09	-0.9200
25	2.24	0.4700
26	1.39	0.1700
27	2.81	0.7600
28	2.06	0.2800
29	2.84	0.2700
30	0	0
31	0	0
32	0	0
33	1.12	0
34	0	0
35	0	0
36	1.02	-0.1946
37	0	0
38	0	0
39	2.67	0.1260
40	0.6563	0.2353
41	10.00	2.5000
42	11.50	2.5000
43	0	0
44	2.6755	0.0484
45	2.08	0.2100
46	1.507	0.2850
47	2.0312	0.3259
48	2.412	0.0220
49	1.64	0.2900
50	1.00	-1.4700
51	3.37	-1.2200
52	1.58	0.3000
53	2.527	1.1856
54	0	0
55	3.22	0.0200
56	2.00	0.7360
57	0	0
58	0	0
59	2.34	0.8400
60	2.088	0.7080

Table A.2 (continued)
Load bus data

Bus number	Real load (pu)	Reactive load (pu)
61	1.04	1.2500
62	0	0
63	0	0
64	0.09	0.8800
65	0	0
66	0	0
67	3.20	1.5300
68	3.29	0.3200
69	0	0

Table A.3. Line data

From Bus	To Bus	Resistance (pu)	Reactance (pu)	Line charging (pu)	Tap ratio
54	1	0	0.0181	0	1.0250
58	2	0	0.0250	0	1.0700
62	3	0	0.0200	0	1.0700
19	4	0.0007	0.0142	0	1.0700
20	5	0.0009	0.0180	0	1.0090
22	6	0	0.0143	0	1.0250
23	7	0.0005	0.0272	0	0
25	8	0.0006	0.0232	0	1.0250
29	9	0.0008	0.0156	0	1.0250
31	10	0	0.0260	0	1.0400
32	11	0	0.0130	0	1.0400
36	12	0	0.0075	0	1.0400
17	13	0	0.0033	0	1.0400
41	14	0	0.0015	0	1.0000
42	15	0	0.0015	0	1.0000
18	16	0	0.0030	0	1.0000
36	17	0.0005	0.0045	0.3200	0
49	18	0.0076	0.1141	1.1600	0
68	19	0.0016	0.0195	0.3040	0
19	20	0.0007	0.0138	0	1.0600
68	21	0.0008	0.0135	0.2548	0
21	22	0.0008	0.0140	0.2565	0
22	23	0.0006	0.0096	0.1846	0
23	24	0.0022	0.0350	0.3610	0
68	24	0.0003	0.0059	0.0680	0
54	25	0.0070	0.0086	0.1460	0

Table A.3 (continued)
Line data

From Bus	To Bus	Resistance (pu)	Reactance (pu)	Line charging (pu)	Tap ratio
25	26	0.0032	0.0323	0.5310	0
37	27	0.0013	0.0173	0.3216	0
26	27	0.0014	0.0147	0.2396	0
26	28	0.0043	0.0474	0.7802	0
26	29	0.0057	0.0625	1.0290	0
28	29	0.0014	0.0151	0.2490	0
53	30	0.0008	0.0074	0.4800	0
61	30	0.0019	0.0183	0.2900	0
61	30	0.0019	0.0183	0.2900	0
30	31	0.0013	0.0187	0.3330	0
53	31	0.0016	0.0163	0.2500	0
30	32	0.0024	0.0288	0.4880	0
32	33	0.0008	0.0099	0.1680	0
33	34	0.0011	0.0157	0.2020	0
35	34	0.0001	0.0074	0	0.9460
34	36	0.0033	0.0111	1.4500	0
61	36	0.0022	0.0196	0.3400	0
61	36	0.0022	0.0196	0.3400	0
68	37	0.0007	0.0089	0.1342	0
31	38	0.0011	0.0147	0.2470	0
33	38	0.0036	0.0444	0.6930	0
41	40	0.0060	0.0840	3.1500	0
48	40	0.0020	0.0220	1.2800	0
42	41	0.0040	0.0600	2.2500	0
18	42	0.0040	0.0600	2.2500	0
17	43	0.0005	0.0276	0	0
39	44	0	0.0411	0	0
43	44	0.0001	0.0011	0	0
35	45	0.0007	0.0175	1.3900	0
39	45	0	0.0839	0	0
44	45	0.0025	0.0730	0	0
38	46	0.0022	0.0284	0.4300	0
53	47	0.0013	0.0188	1.3100	0
47	48	0.0025	0.0268	0.4000	0
47	48	0.0025	0.0268	0.4000	0
46	49	0.0018	0.0274	0.2700	0
45	51	0.0004	0.0105	0.7200	0
50	51	0.0009	0.0221	1.6200	0
37	52	0.0007	0.0082	0.1319	0
55	52	0.0011	0.0133	0.2138	0
53	54	0.0035	0.0411	0.6987	0
54	55	0.0013	0.0151	0.2572	0

Table A.3 (continued)
Line data

From Bus	To Bus	Resistance (pu)	Reactance (pu)	Line charging (pu)	Tap ratio
55	56	0.0013	0.0213	0.2214	0
56	57	0.0008	0.0128	0.1342	0
57	58	0.0002	0.0026	0.0434	0
58	59	0.0006	0.0092	0.1130	0
57	60	0.0008	0.0112	0.1476	0
59	60	0.0004	0.0046	0.0780	0
60	61	0.0023	0.0363	0.3804	0
58	63	0.0007	0.0082	0.1389	0
62	63	0.0004	0.0043	0.0729	0
64	63	0.0016	0.0435	0	1.0600
62	65	0.0004	0.0043	0.0729	0
64	65	0.0016	0.0435	0	1.0600
56	66	0.0008	0.0129	0.1382	0
65	66	0.0009	0.0101	0.1723	0
66	67	0.0018	0.0217	0.3660	0
67	68	0.0009	0.0094	0.1710	0
53	27	0.0320	0.3200	0.4100	1.0000
69	18	0.0006	0.0144	1.0300	0
50	69	0.0006	0.0144	1.0300	0

Appendix B
16-machine, 5-area System Dynamic Data

Table B.1. Machine data

Machine	Bus	Base MVA	X_{ls} (pu)	R_s (pu)	X_d (pu)	X_d' (pu)	X_d'' (pu)	T_{do}' (sec)	T_{do}'' (sec)
1	1	100	0.0125	0.0	0.1	0.031	0.025	10.2	0.05
2	2	100	0.035	0.0	0.295	0.0697	0.05	6.56	0.05
3	3	100	0.0304	0.0	0.2495	0.0531	0.045	5.7	0.05
4	4	100	0.0295	0.0	0.262	0.0436	0.035	5.69	0.05
5	5	100	0.027	0.0	0.33	0.066	0.05	5.4	0.05
6	6	100	0.0224	0.0	0.254	0.05	0.04	7.3	0.05
7	7	100	0.0322	0.0	0.295	0.049	0.04	5.66	0.05
8	8	100	0.028	0.0	0.29	0.057	0.045	6.7	0.05
9	9	100	0.0298	0.0	0.2106	0.057	0.045	4.79	0.05
10	10	100	0.0199	0.0	0.169	0.0457	0.04	9.37	0.05
11	11	100	0.0103	0.0	0.128	0.018	0.012	4.1	0.05
12	12	100	0.022	0.0	0.101	0.031	0.025	7.4	0.05
13	13	200	0.0030	0.0	0.0296	0.0055	0.004	5.9	0.05
14	14	100	0.0017	0.0	0.018	0.00285	0.0023	4.1	0.05
15	15	100	0.0017	0.0	0.018	0.00285	0.0023	4.1	0.05
16	16	200	0.0041	0.0	0.0356	0.0071	0.0055	7.8	0.05

Table B.1 (continued)
Machine data

Machine	X_q (pu)	X_q' (pu)	X_q'' (pu)	T_{qo}' (sec)	T_{qo}'' (sec)	H (sec)	D
1	0.069	0.028	0.025	1.5	0.035	42.0	4.0
2	0.282	0.060	0.05	1.5	0.035	30.2	9.75
3	0.237	0.050	0.045	1.5	0.035	35.8	10
4	0.258	0.040	0.035	1.5	0.035	28.6	10
5	0.31	0.060	0.05	0.44	0.035	26.0	3
6	0.241	0.045	0.04	0.4	0.035	34.8	10
7	0.292	0.045	0.04	1.5	0.035	26.4	8
8	0.280	0.050	0.045	0.41	0.035	24.3	9
9	0.205	0.050	0.045	1.96	0.035	34.5	14
10	0.115	0.045	0.04	1.5	0.035	31.0	5.56
11	0.123	0.015	0.012	1.5	0.035	28.2	13.6
12	0.095	0.028	0.025	1.5	0.035	92.3	13.5
13	0.0286	0.005	0.004	1.5	0.035	248.0	33
14	0.0173	0.0025	0.0023	1.5	0.035	300.0	100
15	0.0173	0.0025	0.0023	1.5	0.035	300.0	100
16	0.0334	0.006	0.0055	1.5	0.035	225.0	50

Table B.2. DC excitation system data

Machine no.	T_r (sec)	K_A	T_A (sec)	V_{rmax} (pu)	V_{rmin} (pu)	K_E	T_E (sec)	A_{ex}	B_{ex}
1	0.01	40	0.02	10	-10	1	0.785	0.07	0.91
2	0.01	40	0.02	10	-10	1	0.785	0.07	0.91
3	0.01	40	0.02	10	-10	1	0.785	0.07	0.91
4	0.01	40	0.02	10	-10	1	0.785	0.07	0.91
5	0.01	40	0.02	10	-10	1	0.785	0.07	0.91
6	0.01	40	0.02	10	-10	1	0.785	0.07	0.91
7	0.01	40	0.02	10	-10	1	0.785	0.07	0.91
8	0.01	40	0.02	10	-10	1	0.785	0.07	0.91

Table B.3. Static excitation system and PSS data

Machine	T_r (sec)	K_a	V_{rmax} (pu)	V_{rmin} (pu)	K_{pss}	T_1 (sec)	T_2 (sec)	T_3 (sec)	T_4 (sec)
9	0.01	200	5	-5	$\frac{12}{377}$	0.1	0.2	0.1	0.2

Appendix C
Jacobian of the FACTS Power Injection

C.1 Thyristor controlled series capacitor (TCSC)
C.1.1 W.r.t state variables

$$\frac{\partial P_k}{\partial k_c} = -\frac{1}{(k_c - 1)^2} V_k V_m B_{km} \sin(\theta_k - \theta_m)$$

$$\frac{\partial Q_k}{\partial k_c} = -\frac{1}{(k_c - 1)^2} B_{km} \left[V_k^2 - V_k V_m \cos(\theta_k - \theta_m)\right]$$

$$\frac{\partial P_m}{\partial k_c} = -\frac{1}{(k_c - 1)^2} V_m V_k B_{mk} \sin(\theta_m - \theta_k)$$

$$\frac{\partial Q_m}{\partial k_c} = -\frac{1}{(k_c - 1)^2} B_{mk} \left[V_m^2 - V_m V_k \cos(\theta_m - \theta_k)\right]$$

C.1.2 W.r.t algebraic variables

$$\frac{\partial P_k}{\partial \theta_k} = \frac{k_c}{(k_c - 1)} V_k V_m B_{km} \cos(\theta_k - \theta_m) \qquad \frac{\partial Q_k}{\partial \theta_k} = \frac{k_c}{(k_c - 1)} V_k V_m B_{km} \sin(\theta_k - \theta_m)$$

$$\frac{\partial P_k}{\partial \theta_m} = -\frac{k_c}{(k_c - 1)} B_{km} V_k V_m \cos(\theta_k - \theta_m) \qquad \frac{\partial Q_k}{\partial \theta_m} = -\frac{k_c}{(k_c - 1)} B_{km} V_k V_m \sin(\theta_k - \theta_m)$$

$$\frac{\partial P_k}{\partial V_k} = \frac{k_c}{(k_c - 1)} V_m B_{km} \sin(\theta_k - \theta_m) \qquad \frac{\partial Q_k}{\partial V_k} = \frac{k_c}{(k_c - 1)} B_{km} \left[2V_k - V_m \cos(\theta_k - \theta_m)\right]$$

$$\frac{\partial P_k}{\partial V_m} = \frac{k_c}{(k_c - 1)} V_k B_{km} \sin(\theta_k - \theta_m) \qquad \frac{\partial Q_k}{\partial V_m} = -\frac{k_c}{(k_c - 1)} V_k B_{km} \cos(\theta_k - \theta_m)$$

$$\frac{\partial P_m}{\partial \theta_k} = -\frac{k_c}{(k_c - 1)} V_m V_k B_{mk} \cos(\theta_m - \theta_k) \qquad \frac{\partial Q_m}{\partial \theta_k} = -\frac{k_c}{(k_c - 1)} V_m V_k B_{mk} \sin(\theta_m - \theta_k)$$

$$\frac{\partial P_m}{\partial \theta_m} = \frac{k_c}{(k_c - 1)} V_m V_k B_{mk} \cos(\theta_m - \theta_k) \qquad \frac{\partial Q_m}{\partial \theta_m} = \frac{k_c}{(k_c - 1)} V_m V_k B_{mk} \sin(\theta_m - \theta_k)$$

$$\frac{\partial P_m}{\partial V_k} = \frac{k_c}{(k_c - 1)} V_m B_{mk} \sin(\theta_m - \theta_k) \qquad \frac{\partial Q_m}{\partial V_k} = -\frac{k_c}{(k_c - 1)} V_m B_{mk} \cos(\theta_m - \theta_k)$$

$$\frac{\partial P_m}{\partial V_m} = \frac{k_c}{(k_c - 1)} V_k B_{mk} \sin(\theta_m - \theta_k) \qquad \frac{\partial Q_m}{\partial V_m} = \frac{k_c}{(k_c - 1)} B_{mk} [2V_m - V_k \cos(\theta_m - \theta_k)]$$

C.2 Static VAr compensator (SVC)
C.2.1 W.r.t state variables

$$\frac{\partial Q_k}{\partial B_{svc}} = V_k^2$$

C.2.2 W.r.t algebraic variables

$$\frac{\partial Q_k}{\partial V_k} = 2V_k B_{svc}$$

C.3 Thyristor controlled phase angle regulator (TCPAR)
C.3.1 W.r.t state variables

$$\frac{\partial P_k}{\partial \phi} = V_k V_m \left[G_{km} \sin(\theta_{km} + \phi) - B_{km} \cos(\theta_{km} + \phi) \right]$$

$$\frac{\partial Q_k}{\partial \phi} = -V_k V_m \left[G_{km} \cos(\theta_{km} + \phi) + B_{km} \sin(\theta_{km} + \phi) \right]$$

$$\frac{\partial P_m}{\partial \phi} = -V_m V_k \left[G_{mk} \sin(\theta_{mk} - \phi) - B_{mk} \cos(\theta_{mk} - \phi) \right]$$

$$\frac{\partial Q_m}{\partial \phi} = V_m V_k \left[G_{mk} \cos(\theta_{mk} - \phi) + B_{mk} \sin(\theta_{mk} - \phi) \right]$$

C.3.2 W.r.t algebraic variables

$$\frac{\partial P_k}{\partial \theta_k} = -V_k V_m \left[G_{km} \left\{ \sin \theta_{km} - \sin (\theta_{km} + \phi) \right\} - B_{km} \left\{ \cos \theta_{km} - \cos (\theta_{km} + \phi) \right\} \right]$$

$$\frac{\partial P_k}{\partial \theta_m} = V_k V_m \left[G_{km} \left\{ \sin \theta_{km} - \sin (\theta_{km} + \phi) \right\} - B_{km} \left\{ \cos \theta_{km} - \cos (\theta_{km} + \phi) \right\} \right]$$

$$\frac{\partial P_k}{\partial V_k} = V_m \left[G_{km} \left\{ \cos \theta_{km} - \cos (\theta_{km} + \phi) \right\} + B_{km} \left\{ \sin \theta_{km} - \sin (\theta_{km} + \phi) \right\} \right]$$

$$\frac{\partial P_k}{\partial V_m} = V_k \left[G_{km} \left\{ \cos \theta_{km} - \cos (\theta_{km} + \phi) \right\} + B_{km} \left\{ \sin \theta_{km} - \sin (\theta_{km} + \phi) \right\} \right]$$

$$\frac{\partial Q_k}{\partial \theta_k} = V_k V_m \left[G_{km} \left\{ \cos \theta_{km} - \cos (\theta_{km} + \phi) \right\} + B_{km} \left\{ \sin \theta_{km} - \sin (\theta_{km} + \phi) \right\} \right]$$

$$\frac{\partial Q_k}{\partial \theta_m} = -V_k V_m \left[G_{km} \left\{ \cos \theta_{km} - \cos (\theta_{km} + \phi) \right\} + B_{km} \left\{ \sin \theta_{km} - \sin (\theta_{km} + \phi) \right\} \right]$$

$$\frac{\partial Q_k}{\partial V_k} = V_m \left[G_{km} \left\{ \sin \theta_{km} - \sin (\theta_{km} + \phi) \right\} - B_{km} \left\{ \cos \theta_{km} - \cos (\theta_{km} + \phi) \right\} \right]$$

$$\frac{\partial Q_k}{\partial V_m} = V_k \left[G_{km} \left\{ \sin \theta_{km} - \sin (\theta_{km} + \phi) \right\} - B_{km} \left\{ \cos \theta_{km} - \cos (\theta_{km} + \phi) \right\} \right]$$

$$\frac{\partial P_m}{\partial \theta_k} = -V_m V_k \left[G_{mk} \left\{ \sin \theta_{mk} - \sin (\theta_{mk} - \phi) \right\} - B_{mk} \left\{ \cos \theta_{mk} - \cos (\theta_{mk} - \phi) \right\} \right]$$

$$\frac{\partial P_m}{\partial \theta_m} = V_m V_k \left[G_{mk} \left\{ \sin \theta_{mk} - \sin (\theta_{mk} - \phi) \right\} - B_{mk} \left\{ \cos \theta_{mk} - \cos (\theta_{mk} - \phi) \right\} \right]$$

$$\frac{\partial P_m}{\partial V_k} = V_m \left[G_{mk} \left\{ \cos \theta_{mk} - \cos (\theta_{mk} - \phi) \right\} + B_{mk} \left\{ \sin \theta_{mk} - \sin (\theta_{mk} - \phi) \right\} \right]$$

$$\frac{\partial P_m}{\partial V_m} = V_k \left[G_{mk} \left\{ \cos \theta_{mk} - \cos (\theta_{mk} - \phi) \right\} + B_{mk} \left\{ \sin \theta_{mk} - \sin (\theta_{mk} - \phi) \right\} \right]$$

$$\frac{\partial Q_m}{\partial \theta_k} = -V_m V_k \left[G_{mk} \left\{ \cos \theta_{mk} - \cos (\theta_{mk} - \phi) \right\} + B_{mk} \left\{ \sin \theta_{mk} - \sin (\theta_{mk} - \phi) \right\} \right]$$

$$\frac{\partial Q_m}{\partial \theta_m} = V_m V_k \left[G_{mk} \left\{ \cos \theta_{mk} - \cos (\theta_{mk} - \phi) \right\} + B_{mk} \left\{ \sin \theta_{mk} - \sin (\theta_{mk} - \phi) \right\} \right]$$

$$\frac{\partial Q_m}{\partial V_k} = V_m \left[G_{mk} \left\{ \sin \theta_{mk} - \sin (\theta_{mk} - \phi) \right\} - B_{mk} \left\{ \cos \theta_{mk} - \cos (\theta_{mk} - \phi) \right\} \right]$$

$$\frac{\partial Q_m}{\partial V_m} = V_k \left[G_{mk} \left\{ \sin \theta_{mk} - \sin (\theta_{mk} - \phi) \right\} - B_{mk} \left\{ \cos \theta_{mk} - \cos (\theta_{mk} - \phi) \right\} \right]$$

Appendix D
Matlab Routine for Controller Design Using LMI Control Toolbox

```
function[K, R, S] = LMIbuilddesign(system, theta, nin, nout, tol);
% Inputs
% system : Generalized regulator
% theta : Angle of conic sector for pole-placement
% nin : Number of control inputs to the system
% nout : Number of measured outputs from the system
% tol : Tolerance

% Outputs
% K : Designed controller
% R, S : Solutions of the LMI solvability conditions
% gopt : Optimum value of gamma

[A, B1, B2, C1, C2, D11, D12, D21, D22] = hinfpar(system, [nout nin]);

n = size(A, 1);

% Inner angle of the conic sector pole-placement region
st = sin(theta); ct = cos(theta);

Ast = A * st; Act = A * ct; B2st = B2 * st; B2ct = B2 * ct; C2st = C2 * st; C2ct = C2 * ct;

% Building the LMIs
setlmis([ ]);

% Define the solution variables
gamma = lmivar(1, [1 0]);
R = lmivar(1, [n 1]);
S = lmivar(1, [n 1]);
Ahat = lmivar(2, [n n]);
```

$Bhat = lmivar(2, [n\ nout]);$
$Chat = lmivar(2, [nin\ n]);$

% Set up the LMIs
$lmiterm([1\ 1\ 1\ R], A, 1, s);$ %LMI#1 : $A * R + R * A'$
$lmiterm([1\ 1\ 1\ Ckhat], B2, 1, \text{'s'});$ %LMI#1 : $B2 * Chat + Chat' * B2'$
$lmiterm([1\ 2\ 1\ 0], B1');$ %LMI#1 : $B1'$
$lmiterm([1\ 2\ 2\ gamma], 1, -1);$ %LMI#1 : $-gamma$
$lmiterm([1\ 3\ 1\ Ahat], 1, 1);$ %LMI#1 : $Ahat$
$lmiterm([1\ 3\ 1\ 0], A');$ %LMI#1 : A'
$lmiterm([1\ 3\ 2\ S], 1, B1);$ %LMI#1 : $S * B1$
$lmiterm([1\ 3\ 2\ Bhat], 1, D21);$ %LMI#1 : $Bhat * D21$
$lmiterm([1\ 3\ 3\ S], A', 1, \text{'s'});$ %LMI#1 : $A' * S + S * A$
$lmiterm([1\ 3\ 3\ Bhat], 1, C2, \text{'s'});$ %LMI#1 : $Bhat * C2 + C2' * Bkhat'$
$lmiterm([1\ 4\ 1\ R], C1, 1);$ %LMI#1 : $C1 * R$
$lmiterm([1\ 4\ 1\ Chat], D12, 1);$ %LMI#1 : $D12 * Chat$
$lmiterm([1\ 4\ 2\ 0], D11);$ %LMI#1 : $D11$
$lmiterm([1\ 4\ 3\ 0], C1);$ %LMI#1 : $C1$
$lmiterm([1\ 4\ 4\ gamma], 1, -1);$ %LMI#1 : $-gamma$

$lmiterm([-2\ 1\ 1\ R], 1, 1);$ %LMI#2 : R
$lmiterm([-2\ 2\ 1\ 0], 1);$ %LMI#2 : 1
$lmiterm([-2\ 2\ 2\ S], 1, 1);$ %LMI#2 : S

$lmiterm([3\ 1\ 1\ R], Ast, 1, \text{'s'});$ %LMI#3 : $Ast * R + R * Ast'$
$lmiterm([3\ 1\ 1\ Chat], B2st, 1, \text{'s'});$ %LMI#3 : $B2st * Chat + Chat' * B2st'$
$lmiterm([3\ 2\ 1\ Ahat], 1, st);$ %LMI#3 : $Ahat * st$
$lmiterm([3\ 2\ 1\ 0], Ast');$ %LMI#3 : Ast'
$lmiterm([3\ 2\ 2\ S], 1, Ast, \text{'s'});$ %LMI#3 : $S * Ast + Ast' * S$
$lmiterm([3\ 2\ 2\ Bhat], 1, C2st, \text{'s'});$ %LMI#3 : $Bhat * C2st + C2st' * Bhat'$
$lmiterm([3\ 3\ 1\ R], 1, Act');$ %LMI#3 : $R * Act'$
$lmiterm([3\ 3\ 1\ -Chat], 1, B2ct');$ %LMI#3 : $Chat' * B2ct'$
$lmiterm([3\ 3\ 1\ R], Act, -1);$ %LMI#3 : $-Act * R$
$lmiterm([3\ 3\ 1\ Chat], B2ct, -1);$ %LMI#3 : $-B2ct * Chat$
$lmiterm([3\ 3\ 2\ -Ahat], 1, ct);$ %LMI#3 : $Ahat' * ct$
$lmiterm([3\ 3\ 2\ 0], -Act);$ %LMI#3 : $-Act$
$lmiterm([3\ 3\ 3\ R], Ast, 1, \text{'s'});$ %LMI#3 : $Ast * R + R * Ast'$
$lmiterm([3\ 3\ 3\ Chat], B2st, 1, \text{'s'});$ %LMI#3 : $B2st * Chat + Chat' * B2st'$
$lmiterm([3\ 4\ 1\ Ahat], 1, -ct);$ %LMI#3 : $-Ahat * ct$
$lmiterm([3\ 4\ 1\ 0], Act');$ %LMI#3 : Act'
$lmiterm([3\ 4\ 2\ S], Act', 1);$ %LMI#3 : $Act' * S$
$lmiterm([3\ 4\ 2\ -Bhat], C2ct', 1);$ %LMI#3 : $C2ct' * Bhat'$
$lmiterm([3\ 4\ 2\ S], 1, -Act);$ %LMI#3 : $-S * Act$
$lmiterm([3\ 4\ 2\ Bhat], 1, -C2ct);$ %LMI#3 : $-Bhat * C2ct$
$lmiterm([3\ 4\ 3\ Ahat], 1, st);$ %LMI#3 : $Ahat * st$
$lmiterm([3\ 4\ 3\ 0], Ast');$ %LMI#3 : Ast'
$lmiterm([3\ 4\ 4\ S], 1, Ast, \text{'s'});$ %LMI#3 : $S * Ast + Ast' * S$
$lmiterm([3\ 4\ 4\ Bhat], 1, C2st, \text{'s'});$ %LMI#3 : $Bhat * C2st + C2st' * Bhat'$

Appendix D: Matlab Routine for Controller Design Using LMI Control Toolbox

% System of LMIs
$syslmi = getlmis;$

% Minimize the cost function
% objective = $alphagamma^2 + eps * Trace(R + S)$
$cobj = zeros(decnbr(syslmi), 1); cobj(1) = 1; penalty = 1e - 8;$
$Rdiag = diag(decinfo(syslmi, R));$
$Sdiag = diag(decinfo(syslmi, S));$
$cobj(Rdiag) = penalty * ones(n, 1); cobj(Sdiag) = penalty * ones(n, 1);$
$slow = 5 + 5 * (tol < 1e - 1); opt = [tol, 300, 1e8, slow, 0];$
$target = 1e - 3;$
$[cost, xopt] = mincx(syslmi, cobj, opt, [\,], target);$

% Retrieve the LMI variables form the solution
$gopt = dec2mat(syslmi, xopt, gamma);$
$R = dec2mat(syslmi, xopt, R);$
$S = dec2mat(syslmi, xopt, S);$
$Ahat = dec2mat(syslmi, xopt, Ahat);$
$Bhat = dec2mat(syslmi, xopt, Bhat);$
$Chat = dec2mat(syslmi, xopt, Chat);$

% Determine $Mti (= inv(M'))$ and $Ni (= inv(N))$ from SVD of $MN' = I - R * S$
$[u, sd, v] = svd(eye(n) - R * S);$ % factorize I-RS
$M = u; N = v * sd'; Ni = inv(N); Mti = inv(M'); Mt = M';$

% Retrieve the controller parameters from the transformed variables
$Ck = Chat * Mti;$
$Bk = Ni * Bhat;$
$Ak = Ni * (Ahat - N * Bk * C2 * R - S * B2 * Ck * Mt - S * A * R) * Mti;$

% Controller state space
$K = ss(Ak, Bk, Ck, 0);$

Appendix E
Matlab Routine for Controller Design Using "$hinfmix$" Function

```
% Inputs
% lmireg : Interactive interface for specifying LMI regions
% dkbnd : 0 is used to ensure strictly proper controller
% tol : tolerance
% nin : Number of control inputs to the system
% nout : Number of measured outputs from the system
% obj : 4-entry vector specifying the H2/Hinf objective
% system : Generalized regulator

% Outputs
% K : Designed controller
% R, S : Solutions of the LMI solvability conditions
% gopt : Optimum value of gamma

region = lmireg;
dkbnd = 0.0;
tol = 0.00001;
r = [0 nout nin];
obj = [0 0 1 0];
[gopt, h2opt, K, R, S] = hinfmix(system, r, obj, region, dkbnd, tol);
```

Index

$\mathcal{H}_\infty$, 25
$\mathcal{H}_2$, 26

accelerating power signal, 63
adaptive control, 79
armature reaction, 61
artificial neural network, 73
automatic governor control, 10
automatic voltage regulators(AVR), 9

Balanced truncation, 29
Bayesian approach, 81
bounded real lemma, 119

centralized control, 151
characteristic polynomial, 106, 109
characteristic vector, 106
classical Smith predictor, 152
condition number, 25
conditional probability, 81
conic sector, 120
Control mode oscillations, 7
controllability, 20
controllability and observability grammian, 27
controllability grammian, 29
controllers, 2
convergence factor, 82
convex programming, 73
critical modes, 16

damping circuits, 1
damping control, 22
damping ratio, 16
damping torque, 10
dead-time system, 153
delay, 152
differential and algebraic (DAE) equations, 14
disturbance rejection, 116

eigen value analysis, 83

eigenvalue assignments, 71
eigenvalue sensitivity, 20, 69
eigenvalues, 16
electrical torque, 61
energy functions, 13
evolutionary programming, 73
excitation system, 15

FACTS devices, 2
feedback signal, 21
frequency signal, 62
Fuzzy set theory, 74

gain and phase margin, 34
gain scheduling control, 79
Gaussian, 81
generalized $\mathcal{H}_\infty$ problem, 116
generalized regulator, 117
genetic algorithms, 73
global positioning system (GPS), 151

H-infinity performance index, 154
H-inifinity optimization, 139
Hankel norm, 30
Hankel singular values (HSV), 27
Heffron and Phillips, 60
HSV, 29
hunting, 1

intelligent control, 73
Interarea mode oscillations, 6
interarea oscillations, 1
interior point technique, 73
intraplant mode, 2
Intraplant mode oscillations, 5

Krylov subspace, 30

least-square solution, 110

left eigenvectors, 18
linear matrix inequality (LMI), 115
linear programming, 70
linguistic variables, 74
local mode, 19
Local plant mode oscillations, 5
loop-shaping, 139
Lyapunov equations, 26

maximum singular value, 3, 24
MIMO, 3
min-max problem, 110
minimal realization, 29
minimum damping ratio, 34
mixed-sensitivity, 116, 139
modal controllability, 20
Modal truncation, 28
modal variables, 18
mode shape, 18
model reduction, 28
modified Smith predictor, 152
mu-synthesis, 139
multi-machine, 2
multiple-model adaptive control (MMAC), 80
multivariable, 3
multivariable control, 23

natural frequency, 7
negative damping, 9, 59
normalized coprime factorization, 139
normalized eigenvalue-distance minimization (WNEDM), 105

observability, 20
observability grammian, 29
orthogonal, 21
oscillations, 1, 9, 11, 12, 36–38, 50, 52, 60, 66, 77, 78, 102, 114, 137, 168

Pacific AC inter-tie (PACI), 10
participation factors, 19
performance, 33
phase compensation circuit, 64
phasor measurement units (PMUs), 151
plant parameter matrix, 107
Pole-placement, 71
pole-placement, 116, 140
pole-zero cancellations, 116
positive damping, 16
power system stabilizers, 79
power systems, 2
principal directions, 23
PSS, 2

recursive algorithm, 81

remote signal, 23
Residualization, 28
residue, 20
Ricatti equation, 115
Riccati based design, 116
Riccati equation, 72
right eigenvectors, 17
Robust performance (RP), 34
robust stability, 34
robust stabilization, 140
robustness, 79
rotor windings, 15
Runge-Kutta solver, 91

Schur's method, 30
self tuning control, 79
sensitivity, 21, 116
Sequential Quadratic Programming (SQP), 110
singular values, 23
singular vectors, 24
SISO, 3
siting, 22
small signal stability, 2
SMIB, 2, 6, 59
Smith predictor, 154
stability, 31
stabilizing signal, 62
state space, 15
structured, 34
SVD, 3
synchronizing torque, 59
synchronous generators, 1

time-delayed system, 153
time-delayed systems, 152
torsional filter, 63
torsional interactions, 63
Torsional mode oscillations, 7
Transformer tap-changing controls, 7
transient stability, 1
trapezoidal rule, 112

uncertainty, 34
unified Smith predictor, 152
unstructured, 34

Voltage instability, 2
voltage regulator, 59

washout circuit, 63
weighting filters, 117
wide-area measurement systems (WAMS), 151

zero eigenvalues, 17